Two Manuals Combined into One Book

Rappelling

And

FM 7-85 Ranger Unit Operations

Department of Defense

Pentagon
Publishing

*TC 21-24

TRAINING CIRCULAR
No. 21-24

HEADQUARTERS
DEPARTMENT OF THE ARMY
WASHINGTON, DC, 10 September 1997

RAPPELLING

CONTENTS

DISTRIBUTION RESTRICTION: Approved for public release; distribution is unlimited.

*This publication supersedes TC 21-24, 24 September 1991.

Page

CHAPTER 3. HELICOPTER RAPPELLING

CHAPTER 4. SPECIAL PATROL INFILTRATION/EXFILTRATION SYSTEM

PREFACE

This circular provides basic rappelling techniques to soldiers and leaders for the conduct of rappelling operations. It serves as the primary reference for both resident and nonresident instruction presented to cadets, officer candidates, and both commissioned and noncommissioned officers. This circular also discusses several advanced techniques dealing with infiltration and exfiltration.

Safety is always the most important consideration when conducting training. This circular provides guidelines for commanders to conduct operations safely. Commanders at all levels must analyze the complete training event to determine the degree of risk involved to men and or equipment (Appendix A). After determining the risks, risk reduction options or controls will be integrated into the training activity. These options or controls may range from a safety briefing or providing additional safety resources, to selecting other means of accomplishing the mission.

Rappelling, FRIES, and SPIES operations are inherently dangerous; therefore, the safety notes and considerations presented in this circular are the minimum acceptable standards.

The proponent of this publication (except for Chapter 5) is the US Army Infantry School. Submit changes for improving this publication on DA Form 2028 and forward it to the Commandant, US Army Infantry School, ATTN: ATSH-RB, Fort Benning, Georgia 31905-5593 (e-mail address: IKNERH@BENNING-EMH2.ARMY.MIL).

The proponent for Chapter 5, Fast-Rope Insertion/Extraction System, is the US Army Special Operations Command. Submit changes or comments for Chapter 5 on DA Form 2028 and forward it to the Commander, US Army Special Operations Command, ATTN: AOOP-TRS, Fort Bragg, North Carolina 28307-5000.

Unless this publication states otherwise, masculine nouns and pronouns do not refer exclusively to men.

CHAPTER 1

TOWER RAPPELLING

The introduction to rappelling is taught on a static tower (Figure 1-1). Using the building block approach to training, soldiers systematically progress to more demanding platforms, including taller static towers, and finally to helicopters. The static tower used may vary in size and height from 34 to 90 feet. The concept of learning the basic rappelling techniques before helicopter operations does not vary. (Units may conduct ground training and wall-side tower rappelling without a rappel master. In this case, the unit commander would be responsible for the safety of the rappelling training.)

Figure 1-1. Rappel tower.

Note: Each static rappel tower should resemble the structure or aircraft being trained to. The Army's standard design for building static rappel towers can be obtained through the Plans and Support Branch of the post Engineers, Building 75, Fort Benning, Georgia 31905 (commercial 706 545-3307 or DSN 835-3307). Strict adherence to the design specifications is required for safety considerations. Rappel towers should be inspected annually by post safety or post engineers.

Section I. PERSONNEL

This section discusses the personnel involved in training rappelling and their duties and responsibilities.

1-1. RAPPEL MASTER

The proponent for accreditation, evaluation, and information for Army rappel master courses is Directorate of Operations and Training, US Army Infantry School, ATTN: ATSH-OTT-F, Fort Benning, Georgia 31905-5593.

a. **Duties and Responsibilities.** The rappel master is responsible for the safety of rappellers. He ensures that all equipment (installation, unit, and personal property) is serviceable. He personally supervises the rappelling operation.

b. **Qualification.** Rappel master qualification is awarded only after the successful completion of a TRADOC accredited rappel master course, which includes the following subjects:

- Duties and responsibilities of a rappel master.
- Safety SOP, regulations, and references.
- Construction of a deployment bag.
- Conduct of an equipment rappel off the rappel tower.
- Conduct of a lock-in.
- Talking a rappeller through completion of a rappel.
- Conduct of ground training.
- Inspecting for proper hook-ups.
- Inspecting and maintaining equipment.
- Inspecting and maintaining snaplinks.
- Inspecting and maintaining rappelling gloves.
- Inspecting and maintaining rappel ropes.
- Identifying the rappel capabilities of aircraft used.
- Controlling rappels from UH-1H or UH-60 aircraft.
- Tying knots (square, bowline, half-hitch, Prusik), safety lines, and rappel seats (Swiss seat, Australian seat).
- Inspecting a rappel seat.
- Aircraft rigging for rappelling operations.
- Aircraft command and control.

c. **Proficiency Maintenance.** To remain current, rappel masters must execute their duties in a tactical or training exercise once every six months. If rappel masters do not execute their duties once every six months, they must take a refresher class taught by a current rappel master. The refresher class includes the subjects listed in the rappel master qualifications paragraph (1-1b).

1-2. RAPPEL SAFETY OFFICER

The RSO is an air assault-qualified or ranger-qualified SFC or above who serves as the OIC for the conduct of rappel operations.

a. The RSO is responsible overall for the safety of all rappellers and ensures that all safety precautions are followed.

b. The RSO briefs VIPs, visitors, and inspecting authorities on training, safety requirements, and layout of training areas.

1-3. RAPPEL LANE NCO

a. **Duties and Responsibilities.** Safety is the rappel lane NCO's number one priority. The rappel lane NCO—

- Ensures proper safety procedures are followed.
- Ensures proper hookup once directed to a rope station.
- Issues commands and maintains eye contact with the rappeller at all times.

b. **Qualifications.** The rappel lane NCO holds the rank of corporal or above, is ranger- or air assault-qualified, and is selected by the commander. Each tower rappel lane must have a qualified rappel lane NCO.

c. **Training.** The rappel lane NCO must also train on the following subjects:

- Responsibilities and safety requirements.
- Inspection and maintenance of equipment.
- Identification of satisfactory anchor points.
- Identification of safe and unsafe hookups.
- Establishment of a rappel point.
- Inspection of a rappel seat.
- Coaching techniques.
- Rappelling procedures.
- Emergency procedures.
- Belay control procedures.

d. **Participation.** The rappel lane NCO must participate in at least seven rappel operations: three as a rappeller; two as an assistant rappel lane NCO; and two performing the duties of a rappel lane NCO under the supervision of a qualified rappel master.

e. **Proficiency Maintenance.** If a rappel lane NCO has not conducted his duties within the last six months, he must complete the training listed in paragraph 1-3c under the supervision of a current rappel master.

1-4. RAPPELLER

Rappel qualification requirements apply to the individual rappeller. Participants in tower rappel training must complete the following requirements under the supervision of a rappel master. The unit commander ensures that personnel successfully complete these requirements before beginning aircraft rappel training.

- Identify all rappelling equipment.
- Demonstrate the construction and attachment of the rappel seat and the rappel rope to the seat.
- Identify unsafe attachments, equipment, rope connections, and seat construction.
- Define terms used in rappelling operations.
- Identify knots used in rappel operations.
- Understand and demonstrate rappel commands.
- Demonstrate rappelling positions.
- Demonstrate belaying procedures.
- Exhibit satisfactory performance from a rappel tower of at least 34 feet in height (two rappels with equipment and weapon, two without equipment and weapon). Two rappels are conducted from the free side of the tower (no wall).
- Demonstrate the ability to lock in.

1-5. BELAYER

Belay requirements are a subtask of basic rappel requirements. Soldiers must know how to belay before conducting rappelling training. The belayer—

- Assumes a position at the base of the lane about one pace away from the tower area.
- Ensures that the rappel ropes are even with the ground during tower rappels.
- Loosely holds the rappel rope with both hands so as not to interfere with the rappeller but still be able to stop the rappeller should he fall.
- Immediately stops the rappeller by pulling downward on the rappel ropes if the rappeller shouts "falling" or loses control of his brake hand during descent.
- Does not wear gloves to ensure a firm grip on the rappelling rope.
- Watches the rappeller at all times, and maintains constant voice or visual contact.
- Wears a helmet to prevent injuries from falling debris.

1-6. BELAY SAFETY

The belay safety must be ranger- or air assault-qualified. He ensures belay personnel are performing their duties properly. Rappel training requires one belay safety for each two rappel stations.

Section II. PREOPERATIONS BRIEFINGS AND SAFETY PROCEDURES

The rappel master ensures participants have a basic understanding of requirements and safety procedures before conducting training.

1-7. SAFETY

The following personnel and equipment must be present during static tower training.

- Two military rappel ropes for each rappel station.
- One safety officer.
- One rappel master for each rappel site.
- One rappel lane NCO per rappel station.
- One medic with medical kit and backboard.
- One safety or medical evacuation vehicle with driver.
- One belayer for each rope station. Rappellers alternate stations.
- One belay safety for each two rappel stations (four ropes).

1-8. SAFETY BRIEFING

As in all training, a safety briefing precedes rappel operations. The rappel master briefs all personnel on safety to include the following instructions.

a. Each rappeller ensures loose clothing and equipment are secured.

b. Rappel seats are tied by the soldier and inspected by the rappel master before climbing the tower. Rappel seats are removed upon completion of every rappel, retied, and reinspected by a qualified rappel master or rappel lane NCO before subsequent rappels.

c. Rappellers climb the tower only when directed by the rappel master or rappel lane NCO.

d. Rappellers stay in the center of the tower until instructed to move to a rappel point.

e. No more than three personnel are behind each rappelling point.

f. If using a troop ladder, only three soldiers are on the ladder at one time. Soldiers do not climb the ladder until told to do so by a rappel master.

g. All rappel masters, rappel lane NCOs, instructors, and anyone else standing near the edge of the top of the tower must wear a restraining strap or safety rope. The strap or rope must be attached to an anchor point.

h. No one should lean or sit on the railing or banisters of the tower.

i. No one is allowed within 3 feet of the edge of the tower without being secured.

j. When attaching the rappel rope to the snaplink, rappellers pull the slack towards the anchor point. The rappel master or rappel lane NCO physically check each hookup.

k. All personnel weighing more than 200 pounds will conduct a standard hookup rappel to determine if they require a friction hookup. A friction hookup is created by placing an additional two ropes in the gate of the snaplink (for a total of six ropes in the snaplink).

l. Combat equipment is positioned on the rappeller so that it does not interfere with the brake hand. The weapon must be slung diagonally across the back with the muzzle pointing down, and on the opposite side of the brake hand.

m. Heavy duty gloves are required for all rappel training.

n. Kevlar helmets with chin straps fastened are worn during tower rappel training.

o. While on the tower, the rappeller maintains eye contact with the rappel master or rappel lane NCO and receives all commands from them.

p. The rappeller ensures that he has a belayer on his rope.

q. The belayer does not wear gloves and keeps both hands on the rope at all times. He also faces the rappeller at all times.

r. All tower rappelling is performed with a double strand of rope.

s. No running is allowed on the tower.

t. No smoking or eating is allowed near the tower.

u. All participants who are unable to rappel, lack confidence, or refuse to rappel are reported to the rappel master or OIC. These participants are immediately removed from the training area.

v. The RSO and rappel master must be aware of overconfidence and carelessness of some rappellers. The rappel master ensures all personnel are tower-qualified before beginning aircraft rappel training.

1-9. TOWER SAFETY AND PREPARATION

The rappel master is in charge of the tower. He conducts a visual and physical inspection of every item of equipment to include the structural lumber and timber, the ladder, the platform floor, and all anchor points.

a. The static tower will not be used during thunderstorms or excessively high winds. If ice is present, or if the platform is slick from rain, rappelling will be delayed until conditions are safe.

A B

Figure 1-2. Example of rappel tower anchor points.

b. All rope stations are rigged with two anchor points (Figure 1-2A, B, C). The first anchor point is a middle-of-the-rope knot, and the second is an end-of-the-rope anchor knot. The rappel master removes all the slack between the knots to create equal tension on the anchor points. He ensures that no less than 10 feet of rope is on the ground during static rappelling.

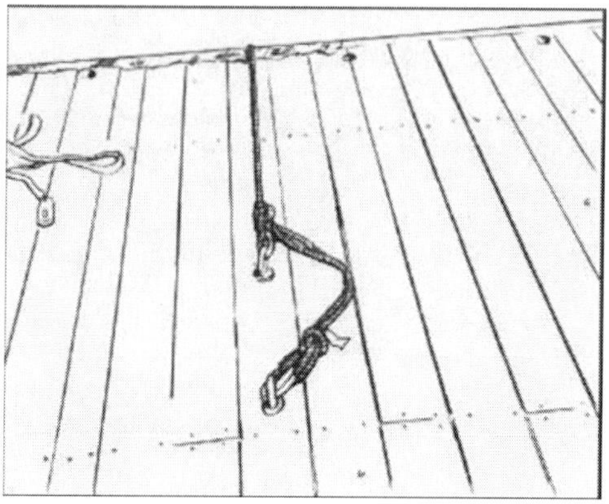

C

Figure 1-2. Example of rappel tower anchor points (continued).

1-10. RAPPELLER PREPARATION

Before conducting a rappel, each rappeller must prepare his individual clothing and equipment.

> **WARNING**
> **Failure to properly prepare rappellers could result in bodily injury or damage to equipment.**

a. Secure shirt tails, loose clothing, equipment, straps, and long hair.

b. Wear a helmet during rappelling. Properly fasten all straps and ensure the helmet is in serviceable condition.

c. Wear heavy leather workman's gloves.

d. Wear identification tags.

e. When rappelling with equipment, LBE or LBV should be unfastened in the front or fastened loosely behind the back of the rappeller. The rucksack should be worn high and tight on the back of the rappeller to allow the brake hand to reach the small of the back. Rucksack adjustment straps will be tied across the chest or tucked away.

f. Sling the weapon diagonally across the back with the muzzle down. Ensure the muzzle is on the guide hand side and the stock is towards the brake hand.

Note: Soldiers rappelling with equipment in excess of 50 pounds may want to consider using a friction hookup.

Section III. RAPPELLING PROCEDURES

This section discusses procedures used in tower rappelling.

1-11. SEAT-HIP RAPPEL

When using the seat-hip rappel, friction is created by a snaplink that is inserted in a sling rope seat and fastened to the rappeller. This method provides a faster and more controlled descent than other methods. Wear gloves to prevent rope burns. An alternate technique is to insert a second snaplink into the first snaplink (attached to rappel seat) and run the rope through the second snaplink. This allows easier disengagement from the rappel rope without running the entire rope through the first snaplink. To disengage from the rappel rope using the alternate technique, release the tension from the rope by opening the gate of the first snaplink and removing the second snaplink (with the rope attached).

a. The rappel seat is constructed as follows (Figure 1-3A through 1-3T):

(1) Place the midpoint (center) of the length of the sling rope on the hip opposite the brake hand (the brake hand is the strong hand) (Figure 1-3A, B, C).

(2) Bring the sling rope around the waist above the hip bone. Tie a double overhand knot over the navel (Figure 1-3D, E, F, G, H).

(3) Let the two free ends of the sling rope fall to the ground in front (Figure 1-3I).

(4) Bring the two free ends of the sling rope down between the legs and up over the buttocks. Ensure that the two free ends do not cross (Figure 1-3J).

(5) Pass the ends of the ropes over the rope that is tied around the waist at the two points above the center of the two rear seat pockets (Figure 1-3K).

(6) Grab the free end of the rope that is on the left side of the body with the left hand, and the free end of the rope that is on the right side of the body with the right hand.

(7) Squat down and simultaneously pull on both running ends of the ropes and stand up. This will tighten the seat.

(8) Take the two running ends of the rope down and back over the waist rope from the inside. Bring the running ends back under the ropes that are going across the buttocks (Figure 1-3L).

(9) Tie the two running ends with a square knot and two overhand knots on the hip opposite the brake hand (Figure 1-3M, N).

(10) Place any excess rope in the trouser pocket near the square knot (Figure 1-3O, P).

(11) With the gate down and the hooked end of the snaplink against the navel, place the end of the snaplink through the single rope that is around the waist and the two ropes forming the double-hand knot (Figure 1-3Q).

(12) Rotate the snaplink a half turn so that the gate is facing up and will open away from the body (Figure 1-3R, S, T).

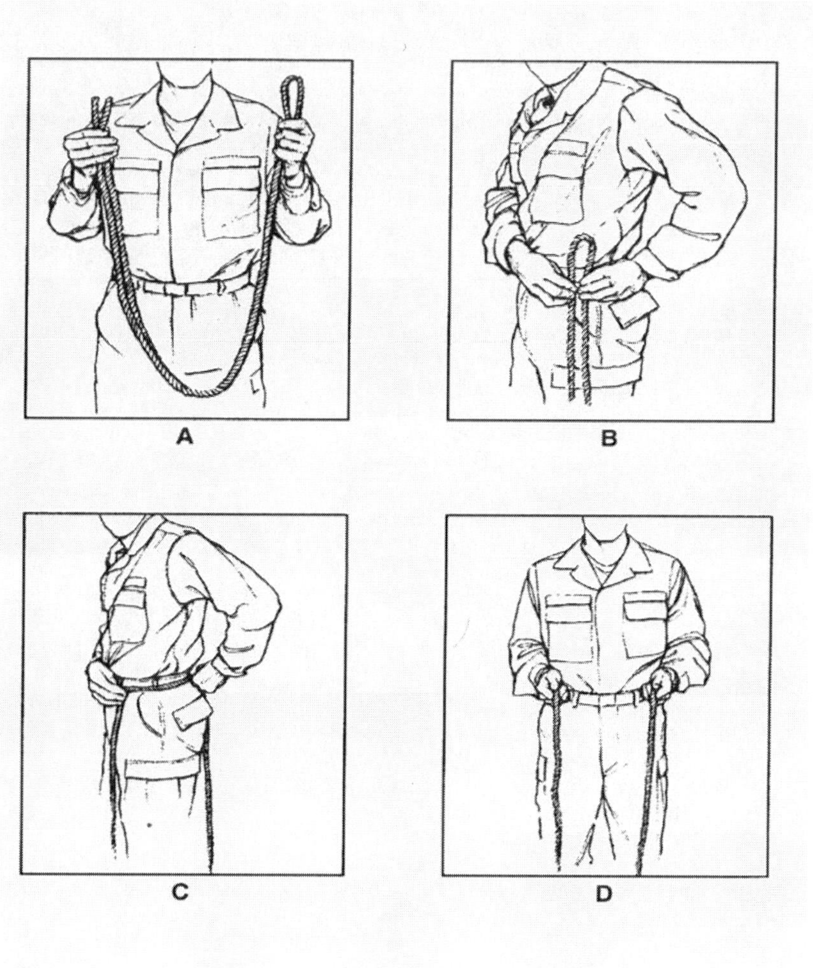

Figure 1-3. Construction of rappel seat.

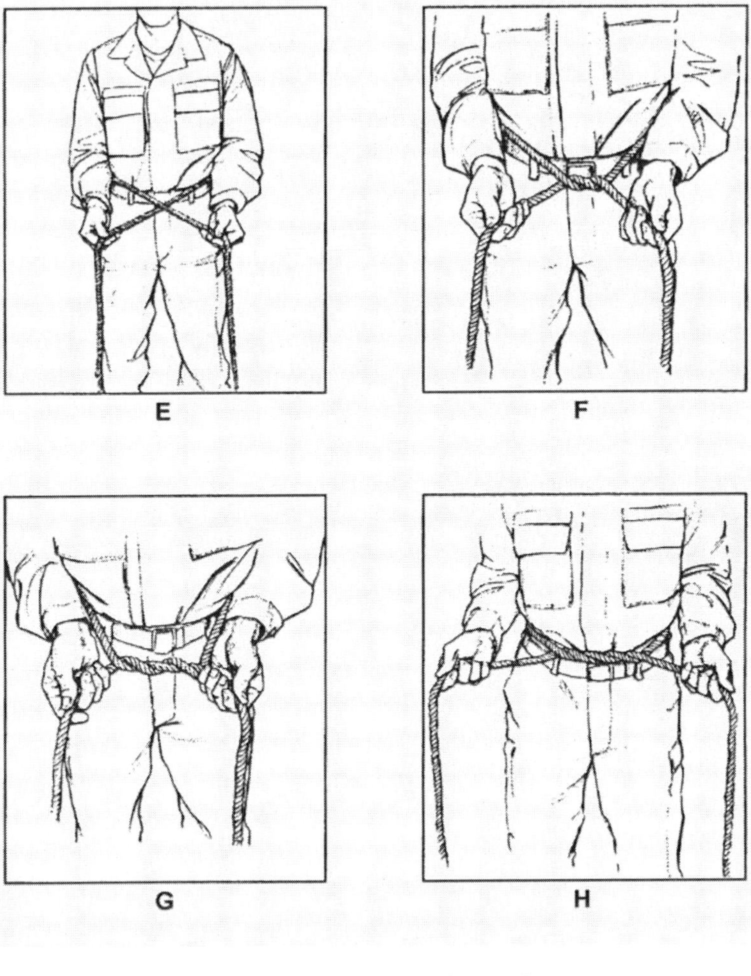

Figure 1-3. Construction of rappel seat.

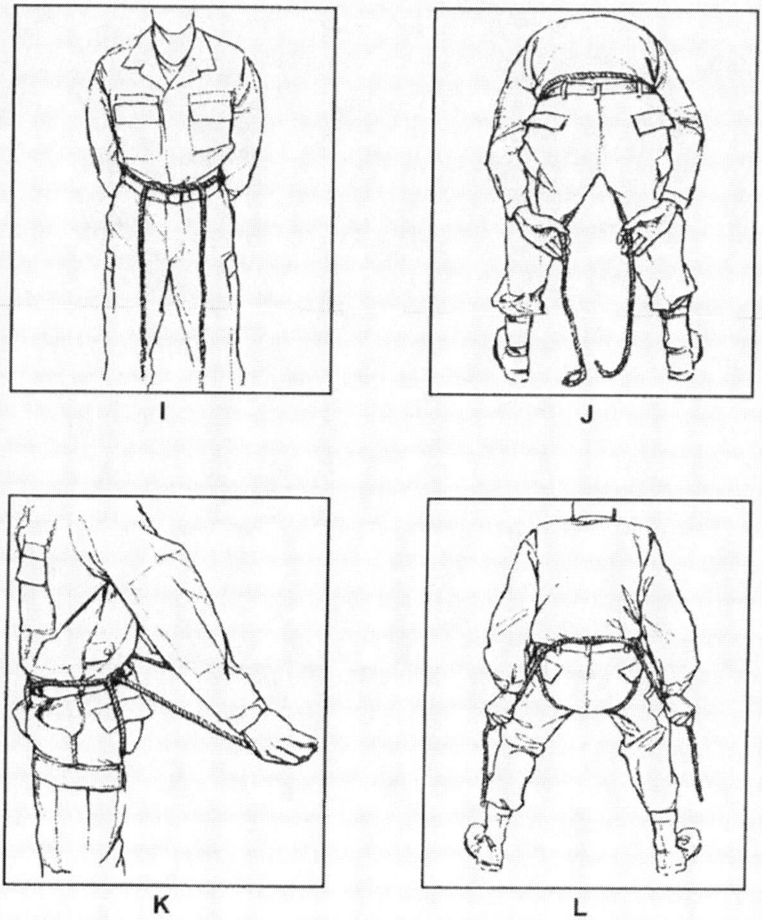

Figure 1-3. Construction of rappel seat.

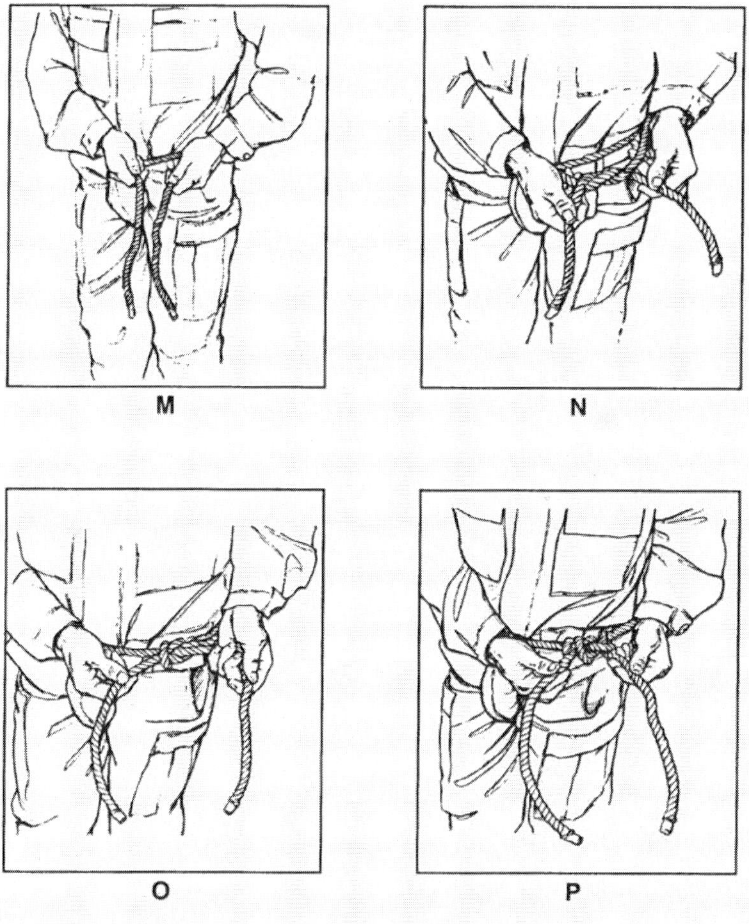

Figure 1-3. Construction of rappel seat.

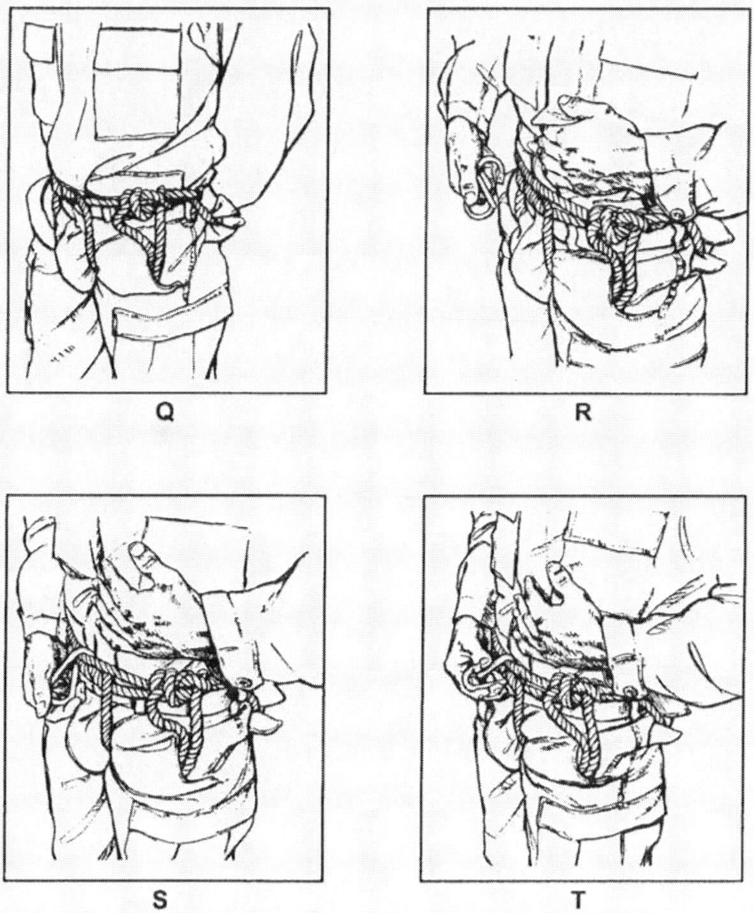

Figure 1-3. Construction of rappel seat.

b. To hook up using the seat-hip method, perform the following:

(1) Place the square knot with two overhand knots towards the anchor point for all seat-hip rappels.

(2) Grasp the two ropes with both hands and drop them through the gate of the snaplink. (At this point, two ropes should be running through the snaplink.)

(3) Using the hand closest to the anchor point, pull the slack towards the anchor point. Rotate the slack under and then over the top of the snaplink.

(4) Drop the two ropes a second time through the gate of the snaplink. (At this point, four ropes should be running through the snaplink.)

(5) Place the guide hand on the rope between the anchor point and the snaplink (palm facing up).

(6) Place the brake hand around the running end of the rope (palm facing down). Place the brake hand with the rope in the small of the back.

1-12. AUSTRALIAN RAPPEL

To hook up for the Australian rappel (Figure 1-4), perform the following:

a. Hook up to the Australian seat.

(1) Grasp the 9-foot sling rope at the midpoint (center) of its length, and double the rope.

(2) Place the doubled sling rope around the back and waist. Ensure that the rope is above the hipbone, but below the ribs.

(3) Tie a *square knot with two overhand knots* over the navel.

(4) Place any excess rope in the trouser pocket nearest the excess.

b. When the Australian seat is donned, face away from the anchor point and to the side of the rappel rope. (Stand to the same side of the rope as the brake hand. Determine left and right of the rope while facing the anchor point.) Place a snaplink onto the seat with the gate facing up, hinge closest to the body. Place it on the hip corresponding to the brake hand. The rappel master grasps the rappel rope and lays the rope into the snaplink. He then places one hand between the snaplink and the anchor point and draws slack toward the anchor point. He rotates the slack down, under, and over the rope and into the snaplink. The rappel master then slides the snaplink directly to the rappellers back.

c. Serve the running end of the rappel rope with the brake hand and prepare to rappel. During descent, brake by drawing the rappel ropes diagonally across the chest with the ropes running from near the waist to the pocket of the opposite shoulder.

Figure 1-4. Australian rappel.

1-13. CLIMBING PROCEDURES

Before climbing the ladder, the rappel master, safety OIC, or rappel lane NCO checks each rappeller's equipment.

a. The rappeller kicks the sand off his boots before climbing.

b. The rappeller grasps the outside of the ladder while climbing, not the rungs (when possible). If stairs are built for the tower, the rappeller grasps the railings as appropriate.

c. Just before climbing up the ladder, the rappeller sounds off, "(name) climbing," and then begins climbing up the ladder. Once at the top and clear of the ladder, the rappeller sounds off, "(name) clear."

d. Once off the ladder, the rappeller waits until the rappel master or lane NCO directs him to proceed to a rope station. At this time, the next rappeller in line may start to climb the ladder.

1-14. TOWER PROCEDURES

After the rappellers climb the tower, the following procedures are adhered to:

a. Once directed to a rope station, the rappel master or lane NCO ensures proper hookup for rappelling.

b. At this time, the rappeller sounds off with "on rappel" and the belayer sounds off with "on belay."

c. While maintaining his brake, the rappeller (on command from the rappel master or lane NCO) steps over the safety rail and faces the anchor point.

d. At this point, the rappel lane NCO sounds off with the following verbal commands and arm-and-hand signals.

(1) *Get Ready*—Rappel master or lane NCO extends both arms to the front with fists clenched and thumbs pointing upward (Figure 1-5). This alerts the rappeller. Each rappeller then looks over his brake hand shoulder to check for the belay man. The rappeller then looks at the rappel master. The rappel master or lane NCO makes his second check of the hookup, rappel seat, snaplink, and equipment.

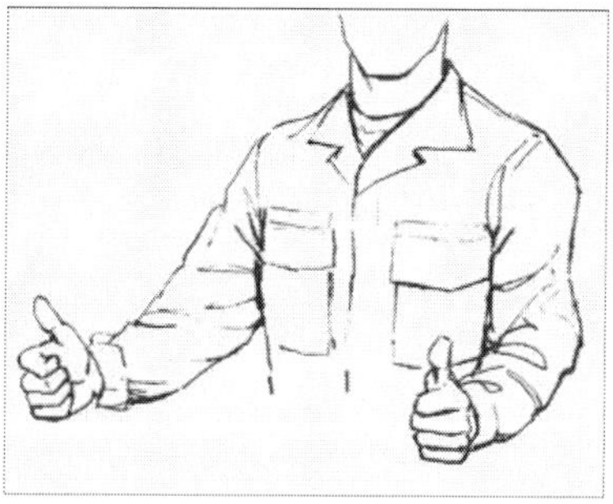

Figure 1-5. Example of arm-and-hand signal GET READY.

(2) *Position*—The lane NCO extends both arms to the front, elbows locked, forearms pointed downward, and fingers extended. He makes a circular motion with both forearms rotating in opposite directions (Figure 1-6). With the brake hand in the small of the back, the rappeller rotates 180 degrees out onto the wall or skid mock up and assumes an L-shaped position. The feet should be shoulder width apart, balls of feet on the wall or skid, knees locked, and body bent at the waist (Figure 1-7).

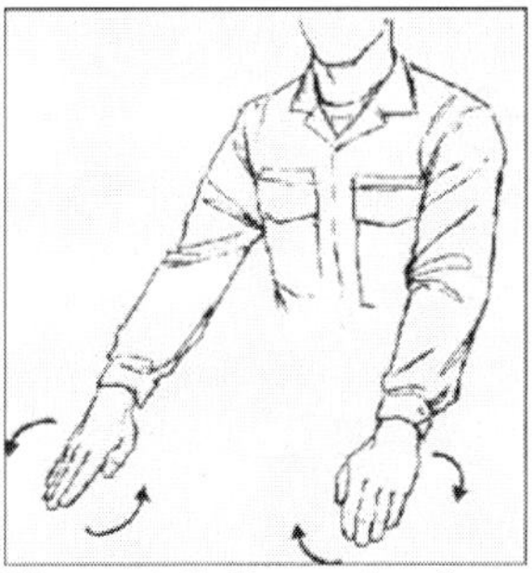

Figure 1-6. Example of arm and hand signal POSITION.

Figure 1-7. L-shape position.

(3) *Go*—The lane NCO extends his right arm with the elbow locked, fingers extended, thumbs around the index fingers, and points directly at the rappeller (Figure 1 8). This initiates the rappel. The rappeller flexes his knees and jumps vigorously backwards. At the same time, the rappeller throws his brake hand out at a 45-degree angle, letting the rope slide through both the brake hand and the guide hand. The rappeller looks over his brake hand shoulder at all times during descent.

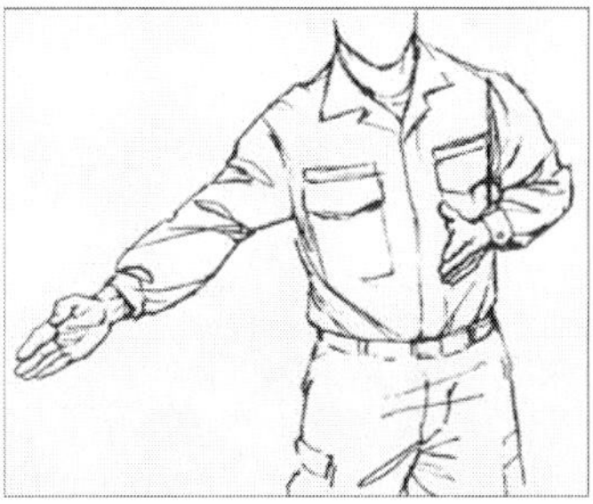

Figure 1-8. Example of arm-and-hand signal GO

e. The rappeller descends in a smooth, controlled manner.

f. The rappeller maintains eye contact with the ground at all times.

g. The rappeller maintains a modified L-shape position during descent with the feet shoulder-width apart, knees flexed, and buttocks parallel to the ground (Figure 1-9).

Figure 1-9. L-shape position while rappelling.

h. When carrying equipment or additional weight, a modified L-shape is used with the legs slightly lower than the buttocks to compensate for the added weight.

i. The rappeller's back is straight. He looks over the brake hand shoulder.

j. The guide hand is extended on the rope with the elbow extended and locked.

k. The rope slides freely through the guide hand, which is used to adjust equipment and to assist in balance during descent.

l. To brake, the rappeller places the brake hand (with rope in hand) in the small of the back and then grasps the rope firmly with the brake hand.

Note: Do not grip the rope firmly with the brake hand while the brake hand and brake arm are extended at the 45-degree angle. If this is done while rappelling, the brake hand and glove may become entangled in the snaplink causing injury to the hand and causing the rappeller to become hung up on the ropes.

m. Releasing tension on the rope and moving the brake hand out to the rear at a 45-degree angle regulates the rate of descent.

n. The rappeller never lets go of the rope with his brake hand until the rappel is completed.

o. After the rope is cleared and the rappeller is off rappel, he acts as the belayer for the next rappeller.

1-15. RAPPEL TOWER TRAINING FOR UH-1H HELICOPTER

Training on the rappel tower for helicopter skid rappelling prepares soldiers to rappel from a UH-1H helicopter (Figure 1-10). The rappeller is hooked up while he sits on the platform just above the helicopter skid. On the rappel master's command GET READY, the rappeller looks over the edge of the tower to ensure the running ends of the ropes are on the ground (Figure 1-10A). On the command SIT IN THE DOOR, the rappeller rotates his feet and legs off the platform and places them on the skid (Figure 1-10B). On the command POSITION, the rappeller turns around and assumes an L-shape position (Figure 1-10C). On the command GO, the rappeller bounds away from the helicopter skid and rappels to the ground. The rappel master is responsible for the proper procedures and safety.

A B C

Figure 1-10. UH-1H helicopter skid rappel training.

1-16. RAPPEL TOWER TRAINING FOR UH-60 BLACKHAWK HELICOPTER

Rappel tower training for the UH-60 Blackhawk is similar to that for the UH-1H with the exception that the UH-60 has no skids on which to stand. Therefore, the edge of the rappel tower is used as a pivot point to assume the L-shape position. All commands are the same except for SIT IN THE DOOR, which does not apply to the UH-60. The stances for each command are also different, as shown in Figure 1-11.

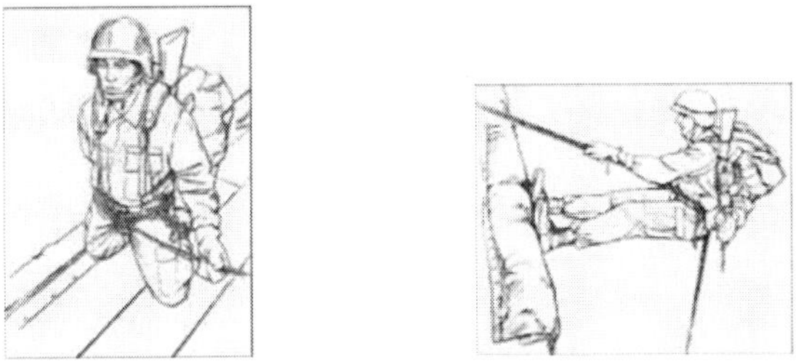

A B

Figure 1-11. UH-60 helicopter rappel training.

1-17. EMERGENCY LOCK-IN PROCEDURES

Mastering the lock-in procedure during tower training is critical before advancing to helicopter rappelling. Using the lock-in procedure allows personnel to hold in position for an extended period of time.

a. If the helicopter gains altitude above the length of the rappel ropes, the rappeller immediately brakes, locks in, and waits for the descent of the aircraft. Procedures for lock-in are as follows:

(1) Place the brake hand in the small of the back and brake to a complete stop.

(2) Release the guide hand from the ropes.

(3) Bring the guide hand around the back and grasp the running end of the two rappel ropes behind the brake hand.

(4) Using the guide hand, bring the two running ends of the rappel ropes around to the front.

(5) Secure these two running ends of the rappel ropes with the two anchor ends of the rappel ropes in the guide hand. This is now the new brake hand.

(6) Take the old brake hand out of the small of the back. Bring it around to the front and grasp the two ropes from the anchor point at a point just above the new brake hand. The old brake hand is now the new guide hand.

(7) Face the rappel master and wait for his command to lower to the ground.

(8) When the command is received from the rappel master to continue the descent, bring the brake hand to a 45-degree angle to the rear. When it is time to brake, bring the new brake hand around to the front diagonally across the chest.

b. If an engine fails or an aircraft emergency occurs during rappelling, the rappellers on the ropes descend as rapidly as possible and move from beneath the aircraft to the sides. Rappellers will maintain control of ropes, if possible.

1-18. COMMUNICATIONS

The rappeller at the top of a rappel point must be able to communicate with those at the bottom. Radios, hand signals, and rope signals are considered during a tactical rappel. For training situations, use the following commands:

COMMAND	GIVEN BY	MEANING
LANE NUMBER, ON RAPPEL	Rappeller	I am ready to begin rappelling.
LANE NUMBER, ON BELAY	Belayer	I am on belay and you may begin your rappel.
LANE NUMBER, OFF RAPPEL	Rappeller	I have completed the rappel, cleared the rappel lane, and am off the rope.
LANE NUMBER, OFF BELAY	Belayer	I am off belay.
LANE NUMBER, FALLING	Rappeller	I am falling. Be alert below—belay man brake.

1-18. DEMONSTRATION

After explaining the procedures to all rappellers, the rappel master should have an assistant demonstrate one complete cycle of rappelling from the static tower. This ensures that the rappellers can hear all the proper commands and see the actions and techniques used on the static tower.

CHAPTER 2
GROUND RAPPELLING

Ground rappelling is a technique that allows soldiers to negotiate mountains and cliffs safely and rapidly. Before rappellers participate in mountain and cliff rappelling they should have completed all the requirements for tower training. To ensure the rappel is safely conducted and meets the mission requirements, rappellers should consider several factors.

2-1. PERSONNEL

The personnel responsibilities for mountain and cliff rappelling are the same as for tower rappelling. As a minimum the following personnel are needed for mountain and cliff rappelling.

a. **Mountaineering Safety Officer (MSO).** The MSO is a SFC or above who is ranger qualified or is a graduate of the Northern Warfare Training Center (NWTC). He has overall responsibility for the safety of all mountaineering participants and ensures that all safety precautions are followed.

b. **Rappel Master**. The rappel master is a qualified officer or NCO in charge of training on ground, tower, and aircraft rappel safety procedures.

c. **Medical Officer or Noncommissioned Officer.** Qualified rappellers and aidmen must know MEDEVAC procedures.

d. **Radiotelephone Operator.** The RATELO monitors the radio and reports to the RSO.

e. **Rappel Lane NCO.** The rappel lane NCO is a qualified officer or NCO in charge of rappelling on a wall or ground rappel lane (Chapter 1, paragraph 1-3).

2-2. SUSTAINMENT TRAINING

Before conducting ground rappel training, the unit or element conducts sustainment training.

a. Review the construction of a rappel seat, seats to be used, and hook-up procedures.

b. Conduct two rappels on the 34-foot tower wall (20-foot tower is acceptable): one with equipment, one without equipment. Conduct two rappels from the free side of a 34-foot or higher rappel tower.

c. Demonstrate the ability to lock in during one of the rappels from the free side of the tower.

2-3. SELECTION OF A RAPPEL POINT

The selection of the rappel point depends on factors such as mission, cover, route, anchor points, and edge composition (loose or jagged rocks). The anchor point should be above the rappeller's departure point. Primary and secondary anchors must be solid (natural anchors are preferred). The rappeller should be sure that the rope reaches the bottom or a place from which he can further rappel or climb. Also, the rappel point should be carefully tested and inspected to ensure the rope will run freely and that the area is clear of obstacles that could be dislodged. If a sling or runner is used for a rappel point, it should be tied twice from two separate loops. Suitable loading and off-loading platforms should be available.

2-4. ESTABLISHMENT OF A RAPPEL POINT

A rappel lane is the area and rope that a rappeller will use during his descent. It should have equal tension between all anchor points by establishing primary and secondary anchor points (Figure 2-1). If one anchor point fails, the rappel rope should not extend. All of the methods discussed below can be performed with a single or double rope. A double rope application should be used for safety.

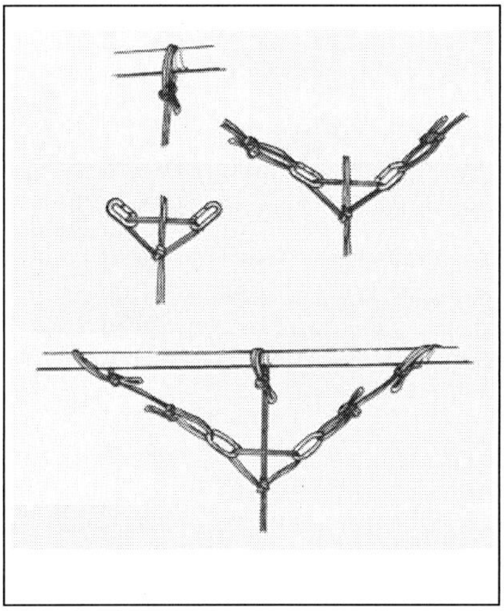

Figure 2-1. Rappel point.

a. If a rappel lane is *less* than half the rope length, the rappeller may apply one of the following techniques:

(1) Double the rope and tie a three-loop bowline around the primary anchor to include the primary anchor inside two loops and enough rope in the third loop to run to the secondary anchor (another three-loop bowline secured with an overhand knot).

(2) Double the rope and tie it around a secure anchor point with a round turn anchor bowline secured with an overhand knot (or any appropriate anchor knot).

(3) Double the rope and establish a self-equalizing anchor system (Figure 2-2) with a bowline-on-a-bight or figure-eight-on-a-bight knot. Tie off on the long standing and with a round turn anchor bowline.

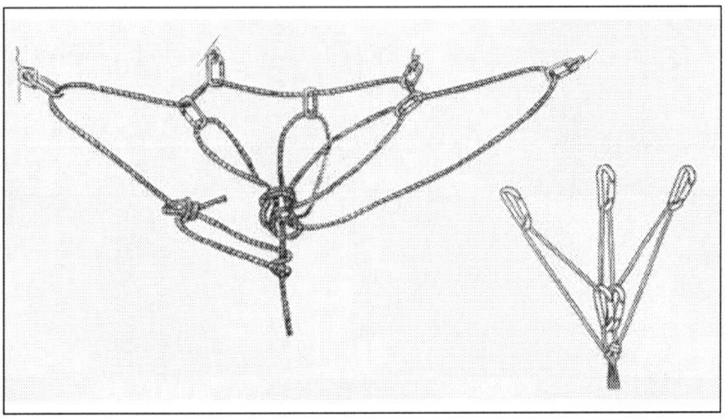

Figure 2-2. Self-equalizing anchor system.

(4) In an emergency, double the rope and place it behind or through a secure anchor point, or tie a runner around an anchor point with a snaplink inserted and place the rope through the snaplink. To preclude a rappeller from sliding off the end of the rappel lane, tie a double figure eight (square knot or double fisherman's knot) at the bottom end of the rope with both ends.

b. If a rappel lane is *greater* than half the rope length, the rappeller may apply one of the following techniques:

(1) Using two ropes, tie a round turn anchor bowline around a primary anchor point. Take the remaining rope (the tail from the primary anchor bowline) and tie another round turn anchor bowline to a secondary anchor point. The secondary anchor point should be in a direct line behind the primary anchor point. The anchor can be either natural or artificial. The ends of the rappel lane ropes should be offset by 15 centimeters (6 inches) so that the rope ends feed freely through the rappeller's snaplink.

(2) Using two ropes, establish a three-piton anchor system using a bowline-on-a-bight knot (or figure-eight-on-a-bight knot) and tie off on the long-standing end with a round turn anchor bowline.

(3) In an emergency, use two ropes and tie the two ends together with a joining knot. Place the joined ropes behind or through an anchor point, or tie a runner around an anchor point with a snaplink inserted and place the joined rope through the snaplink. Offset the joining knot to the left or right of the anchor. Tie off the bottom end of the rope with a joining knot to prevent a rappeller from sliding off the end of the rappel line.

c. Due to the length of the rappel, the rappel rope may not reach the anchor. If the rope is used to tie the knots, it may be too short to accomplish the rappel.

(1) When using a natural anchor—

(a) Tie a sling rope, runner, or another rope around the anchor with a round turn anchor bowline.

(b) Tie a fixed loop (figure eight or butterfly) in one end of the rappel rope, which is attached to the round turn around the anchor through the two snaplinks (opposing gates).

(2) When using an artificial anchor—

(a) Tie off a sling rope, runner, or another rope to form a loop.

(b) Put the loop through the snaplinks that are attached to the artificial anchor point.

(c) Bring the bottom of the loop up and connect it to the snaplinks that are between the artificial anchors (chocks, pitons, or bolts).

(d) Grasp the snaplinks that are between the chocks/pitons and pull them down and together.

(e) Tie a fixed loop (figure eight or butterfly) in the end of the rappel rope and connect this to the snaplinks that have been pulled together.

Notes: 1. Rerouted figure-eight knots can be used instead of bowlines.
2. Runners may be used from one or more anchor points.

d. Backfeed (stack) the rope to ensure it does not snarl when thrown. Take off one wrap at a time and let it fall to the ground, ensuring that no kinks, knots, or twists occur that might hinder the rope from feeding out. When the rope is backfed, anchor off one end of the rope. The two methods used to throw the rope are underhand and overhand. Use the overhand method when trees or shrubs are on or near the rappel point.

e. Set up a retrievable rappel point by applying one of the following techniques:

(1) When the rappel is *less* than half the total length of the rope, double the rope. Place the rope around the primary anchor with the bight formed by the midpoint. Join the tails of the rappel rope and throw the rope over the cliff. Tie a clove hitch knot around a snaplink just below the anchor point ensuring the locking bar inside the snaplink points away from the gate opening end and faces uphill. Snap the opposite standing portion into the snaplink. Upon reaching the bottom of the cliff, pull on the part of the rope to which the snaplink is secured to allow the rope to slide around the anchor point.

(2) When the length of the rappel is *greater* than half the length of the rope used, join two ropes around the anchor point (double fisherman's knot or square knot). Adjust the joining knot so that it is away from the anchor. Tie a clove hitch knot around a snaplink just below the anchor point ensuring the locking bar inside the snaplink points away from the gate opening end and faces uphill. Snap the opposite standing portion into the snaplink. Upon completion of the rappel, pull the rope to which the snaplink is secured to allow the rope to slide around the anchor point.

WARNING

When setting up a retrievable rappel using only a primary point, take care in selecting the point. Ensure soldiers have a safety line when approaching the rappel point, with only the rappeller going near the edge.

2-5. TYPES OF RAPPELS

Types of rappels include body rappel, hasty rappel, seat-hip rappel, seat-shoulder rappel, Australian rappel, and buddy-evacuation. (Seat-hip rappel and Australian rappel are described in Chapter 1, paragraphs 1-10 and 1-11.)

a. **Body Rappel.** Face the anchor point and straddle the rope. Pull the rope from behind and run it around either hip, diagonally across the chest, and back over the opposite shoulder (Figure 2-3). Then, run the rope to the brake hand, which is on the same side of the hip that the rope crosses (for example, the right hip to the left shoulder to the right hand). Lead with the brake hand down and face slightly sideways. The foot corresponding to the brake hand precedes the guide hand at all times. Keep the guide hand on the rope above to guide—not to brake. Lean out at a sharp angle to the rock. Keep the legs spread well apart and relatively straight for lateral stability and the back straight to reduce friction. Turn the BDU collar up to prevent rope burns on the neck. Wear gloves, and use other clothing to pad the shoulders and buttocks. To brake, lean back and face directly toward the rock area so the feet are horizontal to the ground.

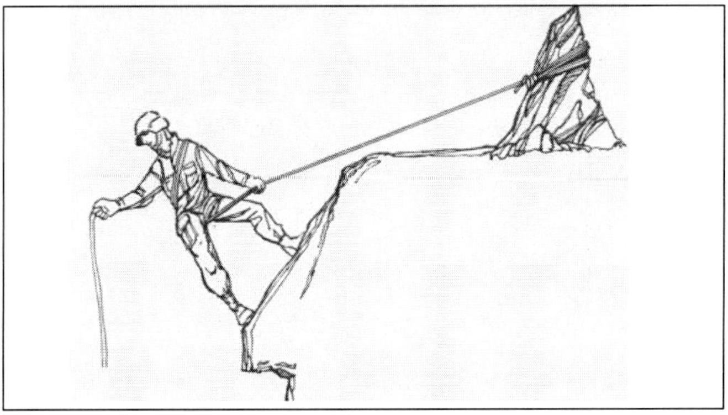

Figure 2-3. Body rappel.

b. **Hasty Rappel.** Facing slightly sideways to the anchor, place the ropes horizontally across the back (Figure 2-4). The hand nearest to the anchor is the guide hand, and the other is the brake hand. Wear gloves to prevent rope burns. To stop, bring the brake hand across in front of the body locking the rope. At the same time, turn to face up toward the anchor point. Use this rappel only on moderate rock pitches. The hasty rappel's main advantage is that it is easier and faster than the other methods, especially when the rope is wet.

Figure 2-4. Hasty rappel.

Figure 2-5. Seat-shoulder rappel.

c. **Seat-Shoulder Rappel.** To hook up for the seat-shoulder method, face the rappel point (Figure 2-5). Snap into the rope that passes up through the snaplink. Bring the rope over one shoulder and back to the opposite hand (left shoulder to right hand). Use the same technique in the descent as in the body rappel. This method is faster than the body rappel, less frictional, and more efficient rappellers with packs and during night operations.

d. **Buddy-Evacuation Rappel.** Use the buddy-evacuation rappel to evacuate an injured soldier from a cliff or steep terrain. Face the cliff and assume a modified L-shape body position to compensate for the weight of the victim on the back. The victim is top-rope belayed from above, which provides the victim with a point of attachment to a secured rope.

Note: To use this rappel, the victim must be conscious.

(1) The method for securing a victim to a rappeller's back is described below:

(a) To secure the victim to the carrier's back with a rope, the carrier ties a standard rappel seat (brake of choice, depending on the injury) and rests his hands on his knees while the victim straddles his back (Figure 2-6.)

(b) A 4.2-meter (14-foot) sling rope is used. A 45-cm (18-inch) tail of the sling is placed on the victim's left hip. (This method describes the procedure for a seat-hip rappel with right-hand brake.)

(c) The long remaining end of the sling rope is routed under the victim's buttocks and passed over the victim's and carrier's right hip. The rope is run diagonally, from right to left, across the carrier's chest, over his left shoulder, and back under the victim's left armpit.

(d) The rope is then run horizontally, from left to right, across the victim's back. The rope is passed under the victim's right armpit and over the carrier's right shoulder.

Figure 2-6. Patient secured to carrier's back.

(e) The rope is run diagonally, from right to left, across the carrier's chest and back across the carrier's and victim's left hip.

(f) The two rope ends should now meet. The two ends are tied together with a square knot and overhand knots.

(g) The knot is positioned on the victim's left hip. The carrier's shoulders should be padded to prevent cutting by the rope.

(2) An alternate method is to use two pistol belts hooked together and draped over the carrier's shoulders. The victim straddles the carrier, and the belay man secures the loose ends of the pistol belts under the victim's buttocks. Slack in the pistol belt's sling should be avoided, since the carrier is most comfortable when the victim rests high on his back. (See FM 8-35.)

(3) A large rucksack can be slit on the sides near the bottom so that the victim can step into it. The victim is belayed from the top with the carrier conducting a standard rappel. The carrier wears the rucksack with the victim inside.

Note: Rucksacks are not designed to support the weight of a soldier and his gear. Therefore, this technique is used only as a last resort.

(4) A casualty secured to a carrier, as described above, can be rappelled down a steep cliff using a seat-shoulder or seat-hip rappel. The casualty's and rappeller's shoulders should be padded where the sling rope and rappel lines cross if a seat-shoulder rappel is used. The buddy team should be belayed from above with a bowline tied around the victim's chest under his armpits. The belay rope must run over the rappeller's guide hand shoulder (Figure 2-7).

Figure 2-7. One-man carry.

2-6. RAPPELLING PROCEDURES

Proper rappelling procedures must be followed to ensure the safety of personnel.

a. The rappel lane NCO is tied off to a separate anchor point for safety, if possible. He—

(1) Ensures that the anchors are sound and the knots are properly tied.

(2) Ensures that loose rock and debris are cleared from the loading platform.

(3) Allows only one man on the loading platform at a time and ensures that the rappel point is orderly run.

(4) Ensures that each man is properly prepared for the particular rapel: gloves on sleeves down, helmet with chin strap fastened, gear prepared properly, and rappe seat and knots correct (if required). He also ensures that the rappeller is correctly hooked up to the rope and is aware of the proper braking position.

(5) Ensures that the proper signals or commands are used.

(6) Dispatches each man down the rope.

(7) Is the last man down the rope.

(8) Ensures that the rope does not run over sharp edges, and pads the rappelling surfaces as necessary to protect the rope.

(9) Ensures that the rope reaches the bottom or is at a place where additional rappels can be made.

b. The first rappeller down—

(1) Selects a smooth route for the rope that is clear of sharp rocks.

(2) Conducts a self-belay with a prusik sling tied from himself to the rappel rope.

(3) Clears the route and places loose rocks, which the rope may dislodge, far enough back on ledges to be out of the way.

(4) Ensures the rope reaches the bottom or is at a place from which additional rappels can be made.

(5) Ensures that the rope runs freely around the rappel point when pulled from below.

(6) Clears the rappel lane by straightening all twists and tangles from the ropes.

(7) Belays subsequent rappellers down the rope.

(8) Takes charge of personnel as they arrive at the bottom (off-loading platform).

Note: A rappeller is always belayed from the bottom except for the first man down. The first man belays himself down the rope by using a safety line attached to his rappel seat that is hooked to the rappel rope with a prusik knot. As the first man rappels down the rope, he "walks" the prusik knot down with him.

2-7. DUTIES OF THE RAPPELLER

To ensure the safe and efficient conduct of rappelling operations, each rappeller must know the duties of his job.

a. Each rappeller down shouts, "Off rappel" (if the tactical situation permits), untangles the ropes, and ensures the ropes run freely around their anchors. When silence is needed, a planned signal of pulling the rope is substituted for the voice signal. After the rope is cleared and the rappeller is off rappel, he acts as the belayer for the next rappeller.

b. All rappellers inspect ropes often when many soldiers are rappelling.

c. The last rappeller to descend constructs a retrievable rappel point and rappels down. Then he gently pulls the rope to prevent the rising rope end from entangling with the other rope. He stands clear of the falling rope and any rocks that it may dislodge.

Note: Soldiers wear gloves for all types of rappels for protection from rope burns. Also, bounding rappels are discouraged since this stresses the anchor and causes undue wear and friction on the rope.

d. Rappellers descend in a smooth, controlled manner. The body forms an L-shape with the feet shoulder-width apart, legs straight, and buttocks parallel to the ground. When carrying equipment or additional weight, a modified L-shape is used with the legs slightly lower than the buttocks to compensate for the added weight. The rappeller's back is straight. He looks over the brake shoulder. The guide hand is extended on the rope with the elbow extended and locked. The rope slides freely through the guide hand, which is used to adjust equipment and to assist balance during the descent. The rappeller grasps the rope firmly with the brake hand and places it in the small of his back. Releasing tension on the rope and moving the brake hand out to his rear at a 45-degree angle regulates the rate of descent. The rappeller never lets go of the rope with his brake hand until the rappel is completed.

e. During training, the lane number must be understood. In combat, a series of tugs on the rope may be substituted for the oral commands to maintain noise discipline. The number of tugs used to indicate each of the commands is IAW the unit SOP.

2-8. BELAYER

The belayer assumes a position at the base of the lane about one pace away from the rock area. He ensures that the rappel ropes are even with the ground during rock and tower rappels. The belayer loosely holds the rappel ropes with both hands so as not to interfere with the rappeller but still able to stop the rappeller should he fall. If the rappeller shouts "Falling" or loses control of his brake hand or descent, the belayer immediately stops the rappeller by pulling downward on the rappel ropes. To ensure a firm grip on the rappelling rope, the belayer does not wear gloves. (Because no friction exists between the belayer's hands and the rappelling rope, gloves are not required for safety. However, whether or not to wear gloves is decided by the OIC/NCO.) The belayer watches the rappeller at all times and maintains constant voice or visual contact with the rappeller. The belayer wears a helmet to prevent injuries from falling debris. All commands are spoken loudly and clearly.

CHAPTER 3

HELICOPTER RAPPELLING

Helicopter rappelling can provide a means of quick insertion with or without an LZ.

Section I. PERSONNEL

The personnel required for helicopter rappelling are the rappel master, the rappel safety officer, the pilot in command, the rappelers, and the belayers.

3-1. RAPPEL MASTER

A qualified rappel master is aboard each aircraft. Safety is the rappel masters number one priority. The rappel master—

- Ensures internal communication between the pilot and rappel master, and external communication between the aircraft and the ground.
- Inspects all equipment and uses only authorized, serviceable equipment.
- Inspects and tests all anchor points and knots before the mission starts.
- Ensures that all rappellers receive a safety briefing and the pilots and aircrew receive an air mission brief.
- Ensures that the rappelers are rappel qualified before conducting helicopter rappelling to include tactical rappelling.
- Maintains communications with the pilot at all times.

3-2. RAPPEL SAFETY OFFICER

The RSO is a SFC or above who is either air assault or ranger qualified. He has overall responsibility for the safety of all rappellers and ensures that all safety precautions are followed. He maintains communications at all times with the pilot and rappel master through FM radio. He alerts the rappel master and pilot of any unsafe acts.

3-3. PILOT IN COMMAND

The pilot in command (PIC) of the aircraft has the following responsibilities:

a. Ensures that the aircrew and all non-aircrew personnel are briefed and understand their responsibilities during rappelling operations, including aircraft safety and action in the event of an emergency.

b. Ensures that the donut ring anchoring device assembly and or aircraft anchor points have been inspected for completeness and functionality, and that they are installed properly.

c. Emphasizes procedural techniques for clearing, recovery, and or jettison of ropes.

d. Keeps the aircraft centered over the target with corrections from the rappel master as required.

3-4. RAPPELLER

In addition to the tower qualification requirements outlined in Chapter 1, the individual rappeller must complete advance training under the supervision of a qualified rappel master in order to participate in tactical helicopter rappelling operations. He must—

- Satisfactorily complete three rappels from a helicopter from a height of 60 feet (two rappels with combat equipment and weapon).
- Demonstrate confidence and proficiency in the techniques, procedures, and equipment used in rappelling from a helicopter.
- Know the rappelling equipment used in helicopter operations and any special equipment required for helicopter rappelling.

3-5. BELAYER

A belayer is assigned to each rope. He is responsible for walking the rope beneath the helicopter during the descent. (Walking the rope is defined as removing the slack from underneath the helicopter by walking backwards with the rope as the helicopter descends to land.) The belayer makes sure the ropes are not caught on the aircraft skids or tires to ensure a safe landing.

Section II. TRAINING

Training for helicopter rappelling includes sustainment training and refresher training.

3-6. SUSTAINMENT TRAINING

Before conducting helicopter rappel training, the unit or element conducts sustainment training.

a. Review the construction of a rappel seat, equipment to be used, and hook-up procedures.

b. Conduct two rappels on the 34-foot or higher tower wall: one without equipment, one with equipment. Conduct two rappels from the open side of a 34-foot or higher rappel tower.

3-7. REFRESHER TRAINING

Refresher training is routinely conducted to maintain acquired skills. Soldiers who have not performed a helicopter rappel during the past six months will undergo refresher training consisting of three satisfactory rappels from a tower (one with weapon and equipment and one executing a lock in) before executing a helicopter rappel.

Section III. PREOPERATIONAL BRIEFINGS AND SAFETY PROCEDURES

This section discusses safety, medical and communications requirements, and the procedures to follow during unusual conditions (adverse weather/terrain conditions, night operations). Personnel must use sound judgement to determine what action to take depending on the nature and severity of the conditions.

3-8. MEDICAL COVERAGE

A qualified and equipped medic will be present to respond to any mishap. Medical transportation must also be available. Absence of a medic, medical equipment, or transportation will terminate training. If the situation warrants and the installation cannot support a MEDEVAC mission, the rappel aircraft may be used as a last resort MEDEVAC vehicle.

3-9. COMMUNICATION REQUIREMENTS
During helicopter rappel training, the RSO will have radio communications with the aircraft. Voice communications are required before starting aircraft rappelling. Additionally, the RSO will inform the PIC to stop operations if an unsafe condition develops.

3-10. ADVERSE WEATHER/TERRAIN CONDITIONS
Rappel operations will not be conducted under the following conditions

- Ambient temperature is 30 degrees Fahrenheit or less.
- Winds in excess of 30 knots.
- Lightning strikes within 1 nautical mile of rappelling operations.
- Wind chill factors caused by the helicopter's rotor wash or extraction cruise air speeds, which could cause cold weather injuries.
- Water or ice on the rope inhibiting the ability of the rappellers to control their descent.
- The rope is exposed to the elements for a sufficient length of time to freeze, thereby reducing its tensile strength.
- Blowing particles produced by rotor wash causes the aircrew or the rappel master to lose visual contact with the ground.

3-11. NIGHT OPERATION REQUIREMENTS
The following requirements are necessary for night rappelling operations.

a. One chemlight will be attached to the end of the rope and one on each rappeller.

b. One chemlight will be secured to the attachment point of the rope.

c. Night vision goggles (NVG) will not be worn by rappellers during the descent. Aircrew members will wear NVG as required during night operations. NVG lighting criteria will be IAW Army regulations, specific aircraft aircrew training manuals, unit SOPs, or the tactical environment.

3-12. SAFETY BRIEFING
The following safety measures are enforced.

a. Loose clothing and equipment are secured.

b. Helmets are worn with chin straps fastened.

c. Rappellers wear identification tags and earplugs, carry identification cards, and role down their sleeves.

d. Weapons are slung diagonally across the back with the muzzle pointing down on the guide hand side.

e. All seats and rappelling equipment must be inspected by a rappel master before rappelling.

f. No running is allowed within 50 feet of the aircraft.

g. Personnel approach and depart the helicopter from the front and forward of the rear of the cargo doors. When approaching or departing the helicopter, personnel bend their bodies forward at the waist to ensure clearance of the rotor blades. At no time will personnel go near the rear of the aircraft.

h. Upon boarding the aircraft, the rappeller sits or kneels down, hooks up, and applies his brake hand to the small of his back.

i. While in the helicopter, the rappeller maintains eye-to-eye contact with the rappel master and receives commands from him.

j. The rappeller ensures that he has a belayer on his rope at all times when conducting training at a hover site.

k. During the descent, the rappeller maintains eye-to-ground contact.

l. If the rappeller sees his rope coming off the ground or sees that his belayer has lost control of his rope, he immediately brakes and executes a lock-in. He then waits for commands from the rappel master.

m. The rappeller brakes once every 30 feet during descent.

n. The belayer does not wear gloves. He keeps both hands on the rope and his eyes on the rappeller at all times.

o. All rappelling will be conducted using a double strand of rope.

Section IV. DEPLOYMENT OF ROPES

Deployment of the ropes from a helicopter is a critical task. It can cause a planned rappelling operation to fail, or it can increase the time required to conduct the operation. This is due to the likelihood of the ropes becoming entangled. To prevent this, ropes must be deployed using a positive control technique. Two of the techniques that may be used are the *deployment bag* technique and the *log coil* technique.

3-13. DEPLOYMENT BAG TECHNIQUE

The rappeller places the deployment bag (D-bag) (standard) on a flat surface with the stow loop facing upward. If the D-bag still has a static line, the rappeller removes it by cutting the static line where it attaches to the bag. A 12-foot section of the static line can be used as a safety line in the helicopter.

a. Use a sandbag (about one E-tool spade full of sand) as a weight in the deployment system. Roll the sandbag into a rectangular shape, tie it, and then place it in a small plastic bag, rolled and taped. Place a retaining band over the middle of the weight, and place the weight on top of the D-bag. The weight should be about the same width as the D-bag.

b. Lay out and inspect the two ropes. The working ends (closest to the D-bag) should be even. Place a round turn with the two working ends of the ropes on the weight's retaining band. Working on top of the deployment bag, start forming figure eights. The stack should consist of 8 to 10 figure eights, one on top of another—do not exceed the width of the D-bag. Then, starting from either side, center a retaining band over the stack. Ensure that it is over all the figure eights in the stack. Repeat the process each time, placing one stack in front of the other. Continue until about 10 feet of rope remains. If one rope is shorter than the other, the end of the shorter rope should be about 10 feet from the last stack.

c. For a primary anchor point, measure down about 4 feet from the end of the shortest rope. Using both ropes, tie a bowline without a half hitch. This knot is the primary anchor point.

d. For a secondary anchor point, tie an end-of-line bowline with a half hitch toward the end of the ropes. The dressed knot should be 18 to 22 inches from the primary anchor point knot when using the UH-1H helicopter and 22 to 30 inches when using the UH-60 helicopter. If excess rope remains, S-fold and tape it between the two knots. Ensure that 3 to 6 feet of rope remain from the last stack of figure eights.

e. To place the ropes in the D-bag, remove the bag from under the stack of rope. Place the weight into the bottom of the D-bag, and place all the stacks of figure eights, in the order they were made, one on top of the other into the bag.

f. Two left, two right, and two rear loops are in the opening of the D-bag. ***Do not use the two rear loops***. Close the flap on the D-bag and push the loops through the aligning holes on the flap. Using an 8-inch piece of gutted 550-pound cord, tie the two loops together with a square knot and two hitches. Ensure that two 120-foot ropes are coming out of the center of the flap—not to one side. Repeat the process and tie the right loops together. Wrap the excess rope lengthwise around the bag.

Note: The sizes of the loops of the bowline should be no larger than an average size fist.

3-14. LOG COIL TECHNIQUE

The rappeller lays the running end of the double rope along the length of the coiling log (Figure 3-1). He then coils the double rope around both the running end of the rope and the coiling log. The rope must be coiled evenly and tightly.

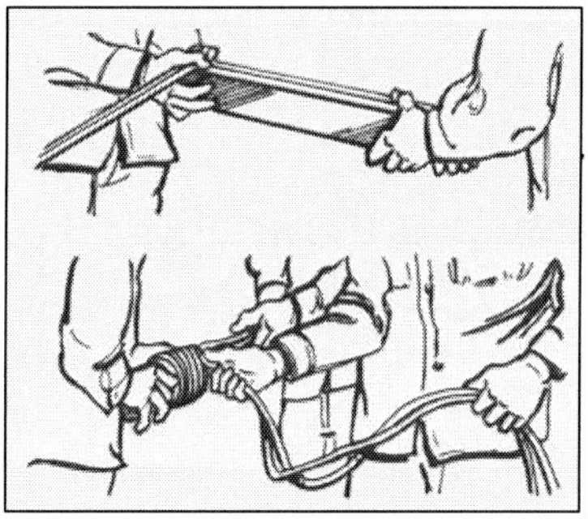

Figure 3-1. Coiling Log

Section V. RAPPELLING OPERATIONS FOR UH-1H IROQUOIS HELICOPTER

The UH-1H helicopter provides a safe, stable aerial platform from which to conduct rappelling operations when landing is not feasible.

3-15. CHARACTERISTICS

The UH-1H is a single-engine, medium-speed, single-main rotor helicopter that can transport eight rappellers, one rappel master, and one three-man crew. It has a lift capacity of 2,300 pounds and is equipped with several floor-mounted tie-down fittings, seven of which are used during rappelling operations.

3-16. RIGGING OF UH-1H HELICOPTER FOR RAPPELLING

The rappel master performs the following to rig the UH-1H helicopter.

a. Removes all seats.

b. Locks the doors in the open position. If no locks are present, removes doors. Also, he removes small cargo doors.

c. Pads and tapes all sharp edges on the floor, door ledge, and all protrusions on the skids. Ensures each door ledge has a scuff pad to protect the rope from contacting the metal door ledge.

d. Secures the donut ring to the center of the floor of the helicopter. The donut ring has six snaphooks numbered clockwise, with 12 o'clock being toward the front of the helicopter. Positions the clamps of the donut ring toward the aft end of the helicopter. Ensures the front two snaplinks and rear two snaplinks are facing outside of the donut ring and hooks them to the tie-down ring with the snaphook facing down. Ensures the center two snaphooks face into the donut ring and are connected to the floor tie-down ring with the gate facing down (Figure 3-2). Attaches the free-floating safety ring to the center floor tie-down ring using a seventh snaphook.

Note: Using this technique has caused equipment failure in many aircraft. The cable clamp nuts must be inspected and certified as airworthy by the pilot or maintenance crew before flying the mission.

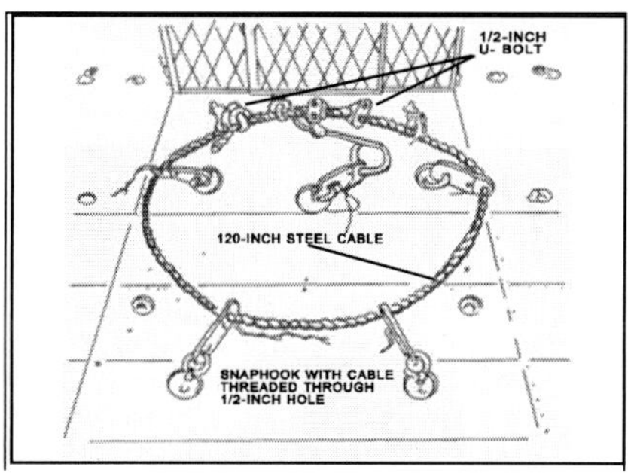

Figure 3-2. Donut ring attached to the floor of the helicopter.

3-17. CONSTRUCTION OF ANCHOR POINTS

In addition to the usual equipment requirements for rappelling operations, an anchor assembly is fabricated. This anchor assembly is commonly known as a *donut ring*. The secondary anchor point is a *floating safety ring*.

a. **Donut Ring.** This primary (or No. 1) anchor point for the rappelling ropes is constructed from a 1/2-inch steel cable with a steel wire core. The cable is 120 inches long, consists of 6 strands (18 wires per strand), and has a tensile strength of 21,000 pounds (Figure 3-2). The completed donut ring has a tensile strength of 3,000 pounds.

(Figure 3-3 shows the donut ring and rappel rope connection in a helicopter.)

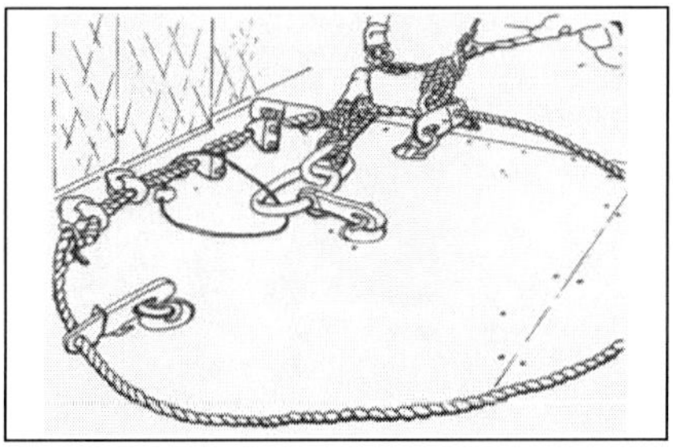

Figure 3-3. Donut ring and rappel rope connection in a helicopter.

(1) Thread six parachute static line snaphooks onto the 120-inch steel cable so that four snaphooks are facing out with the gates down, and the center two snaphooks are facing in with the gates down. Drill the end portion of each snaphook to make a 5/8-inch diameter hole, and thread the cable through the holes in the snaphooks.

(2) Overlap the ends of the 120-inch cable 20 inches to form a circle. Secure the ends with four 1/2-inch U-bolts placed at 2- to 3-inch intervals.

(3) Attach two U-bolts to each dead end of the cable so that the bolts engage the dead end.

(a) Before torquing the U-bolts, position a 12-inch length of chain or 1/8-inch diameter cable on the center of the overlapped 120-inch steel cable so that it remains in position between the two center U-bolts.

(b) Tighten each nut of the U-bolts with a torque wrench (if possible) to 40-foot pounds (480-inch pounds).

(c) After the U-bolt clamps have been attached and tightened, fasten a steel plate (drill to fit) over the open end of the U-bolt studs and spot weld in place to prevent loosening.

b. **Floating Safety Ring**. The floating safety ring is referred to as the secondary (or No. 2) anchor point for the rappelling ropes (Figure 3-4). The snaplink at the end of the rappelling rope is hooked to this connection. Either of the following two types of floating safety rings may be used:

(1) *First method.* Thread an elliptical rappelling ring through the free end of the keeper chain (cable). It is constructed of cold-rolled steel that is 1/4-inch in diameter, with inside dimensions of 2 1/4 inches (minor axis) and 4 inches (major axis). Thread a seventh parachute static line snaphook onto the ring before welding. Weld this ring together so that it withstands a force up to 3,000 pounds.

(2) *Second method.* Attach two snaplinks to the aircraft tie-down ring in the center of the donut ring. Insert the first snaplink through the free end of the keeper chain (cable) and the tie-down ring with the gate down. Insert the second snaplink through the free end of the keeper chain (cable) and the tie-down ring with the gate up. Tape the snaplink

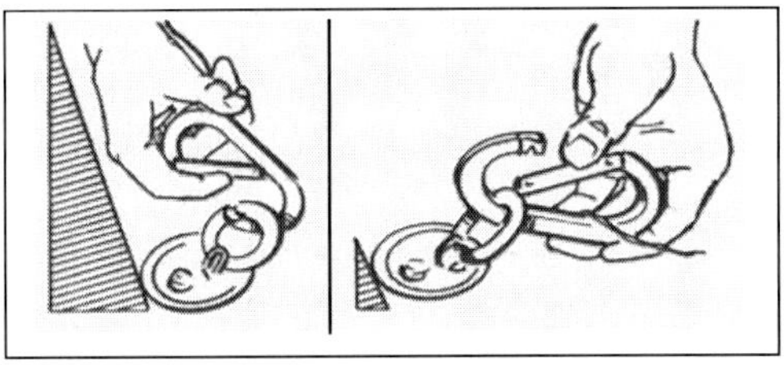

gates closed with masking tape. Then, tape the snaplinks together to ensure that the snaplink gates are on opposite sides of each other.

Figure 3-4. Floating safety ring formed with two snaplinks.

c. **Rappel Rope Anchor Points.** The rappelling rope is connected to the floating safety ring and the donut ring as follows:

(1) *Number 1 anchor point (donut ring).* Using a bight about 5 feet from the end of the standing part of the rope, make one turn through the snaplink forming a round turn. The bowline is the preferred method of attaching rappel ropes to snaplinks or anchor systems. The round turn with two half hitches is a reliable means of attaching ropes to anchor systems. Secure the round turn to the snaplink with two half hitches. Make the connection to the donut ring by attaching the snaplink gate upward. Ensure the gate faces upward with the opening away from the knot.

(2) *Number 2 anchor point (floating ring).* Attach the No. 2 snaplink the same as the first with some exceptions. Using a bight about 2 feet from the end of the standing part of the rope, connect the snaplink to the rope the same as the first connection. Tape the end of the standing part of the rope and the knots with masking tape (or green engineer tape) to secure them in place. Make the connection to the floating safety ring the same as the connection to the donut ring (Figure 3-5). Four ropes can be connected to the floating safety ring using two snaplinks (Figure 3-6).

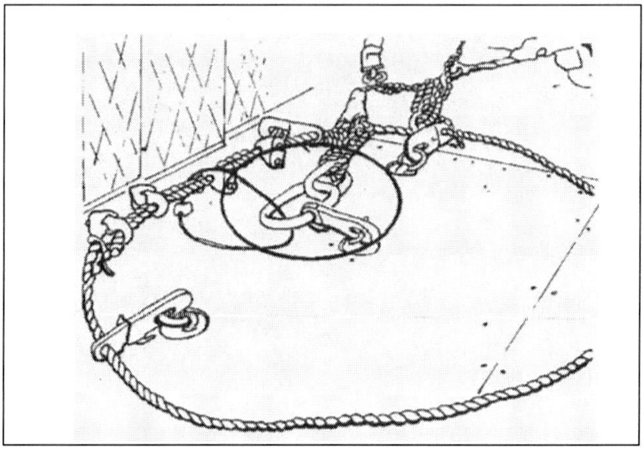

Figure 3-5. Rappel rope connection using two snaplinks for the floating safety ring.

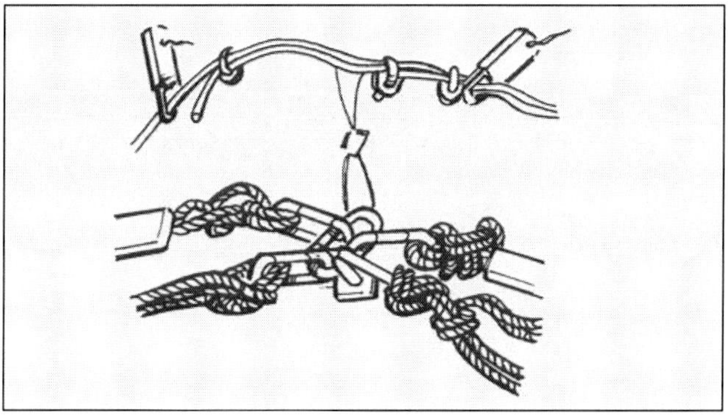

Figure 3-6. Four rappel ropes connected to the floating safety ring (two snaplinks).

3-18. SEATING ARRANGEMENTS AND LOADING TECHNIQUES

Rappelling operations are executed quickly and safely to ensure the success of the operation in a combat or training environment. All rappellers must know the seating arrangements and the procedures for loading (Figure 3-7). They must be rehearsed under the supervision of the rappel master before conducting the operation.

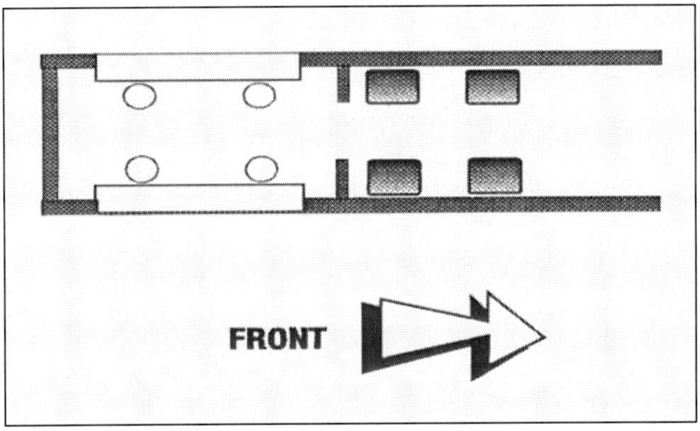

Figure 3-7. Seating arrangement.

a. The first four rappellers are hooked up to the donut ring and use safety belts while in flight. The remaining rappellers use seat belts or safety straps. For safety emphasis during flight, each rappeller's brake hand is in position with no slack in the rope between the brake hand and the donut ring. The rappeller ensures that his rappelling rope is not tangled on any part of the interior of the helicopter or his equipment. He ensures that the coiling log or D-bag is located in such a position that he drops the rope with ease, using his guide hand. The coiling log or D-bag is secured with half of it under the rappeller's inside leg and the other half on top of his outside leg.

b. The rappellers should board the aircraft and sit along the leading edge of each door. They hook up with the rappel seat square knot on the inboard side of the aircraft and the brake hand on the outboard side. The rappel master sits in the center to aid visual inspection of personnel and equipment.

c. For a flight time of less than five minutes and for training and operations where speed is a factor, the rappel master directs the rappellers to be seated in the door facing outward with their feet on the skid. In a tactical situation and before the helicopter comes to a hover, the rappellers assume an upright position facing inside the helicopter.

d. The aircraft is cleared as fast as possible. This reduces the time that the aircraft is in the high-hover attitude to a minimum.

3-19. RAPPELLING PROCEDURES

The rappel master ensures all personnel follow the rappelling procedures for the safe and efficient execution of the operation. These procedures are rehearsed under the supervision of the rappel master.

a. Rappellers (eight maximum) are identified, and rappel masters tell them which door of the aircraft to exit. Rappellers approach the aircraft once it is firmly on the ground, and the crew chief signals when it is safe to approach. The approach to the aircraft should be about 45 degrees to the nose of the aircraft allowing the pilot to see the rappeller.

b. If the rappel master is not already on the aircraft, he boards first and secures himself to the aircraft with a safety harness. He dons a headset to maintain communications with the pilot and positions himself behind the crew chief's chair.

c. As the rappellers approach the aircraft, they hand their anchor attaching points (snaplinks connected to their ropes) to the rappel master. The rappel master connects one set of snaplinks to the floating safety ring and one set of snaplinks (the primary) to the donut ring.

d. All rappellers use safety belts. For safety emphasis during flight, each rappeller's brake hand is in position with no slack in the rope between the brake hand and the donut ring. Rappellers ensure that their rappelling ropes are not tangled on any part of the interior of the helicopter or their equipment. They ensure that they can drop their D-bag or log coil using the guide hand. The D-bag or log coil is placed so it does not fall out of the aircraft. Once over the target area, the rappel master issues the rappel commands and ensures the rappel ropes are on the ground (about 20 feet of rope should be on the ground).

e. On the command GO, the rappeller flexes at the knees and vigorously pushes away from the skid gear, allowing the rope to pass through his brake hand and guide hand. The descent is accomplished smoothly at a rate of about 8 feet per second, avoiding jerky stops. The rappeller initiates his braking action slowly when he is about halfway to the ground. The safety officer or NCO ensures that at least a five-second delay is maintained between each rappeller. Another technique used to deploy rappellers is to issue the command GO to the rappellers diagonally opposite each other. In the early phase of performance, the rappeller's feet are together (particularly if over woods or jungle) and his guide hand is used (not the brake hand) to remove any entanglements.

f. Upon reaching the ground, the rappeller clears the rappel rope through the snaplink (or rappel ring) until the rope is free. If other rappellers are to follow on the same rope, the rappeller on the ground separates or untwists the ropes and becomes the belayer for each subsequent rappeller. If ropes are to be released from the aircraft following the last rappeller, untwisting and separating the ropes is not necessary. A two-second interval is maintained between exit groups (two men exiting at the same time).

g. The rappel master releases the rope from the donut ring after he confirms (by visual inspection) that the rappeller is off rappel. He then drops the rope away from the helicopter.

3-20. RAPPELLING COMMANDS

Helicopter rappelling is conducted in a noise-filled environment. Each rappeller is trained to know the rappel commands and to understand the actions required of him during the execution of each command. Due to the noise created by the helicopter, all rappel commands should be accompanied by arm-and-hand signals. All commands are issued by the rappel master and confirmed orally by the rappellers.

a. The helicopter rappelling commands are as follows:

(1) GET READY. The rappeller checks his combat equipment, looks toward the donut ring, and pulls the rope to check the anchor point connection. He then checks his

snaplink to ensure the rope is properly seated in the snaplink on his rappel seat.

(2) THROW ROPE. The rappeller drops his deployment bag out and away from the helicopter with his guide hand. He ensures that the rope does *not* fall between the side of the helicopter and the skid. The rappel master ensures the rope is touching the ground and is free of tangles and knots. The rappeller then replies, "Rope okay."

(3) SIT IN THE DOOR. The rappeller swings his legs to the outside of the helicopter and takes up a sitting position. (This command applies only to the UH-1H helicopter.)

(4) POSITION. The rappeller pivots 180 degrees on the skid. His feet are shoulder-width apart, the balls of his feet are on the skid, his knees are locked, his body is bent forward in an L-shape position, and his brake hand is on his buttock. The rappel master makes a visual inspection of the snaplink and rappel ring.

(5) GO. The rappel master points at the rappeller to exit. One rappeller from each side rappels at the same time. The rappel master gives the command GO to the rappellers that are diagonal to each other—for example, the front left rappeller and the rear right rappeller. The rappeller flexes his knees and jumps backward, letting the rope run through the brake hand and guide hand.

b. The arm-and-hand signals that are used with rappelling commands are as follows:

(1) GET READY (Figure 3-8). The rappel master extends both arms to the front with fists clenched and thumbs pointing upward.

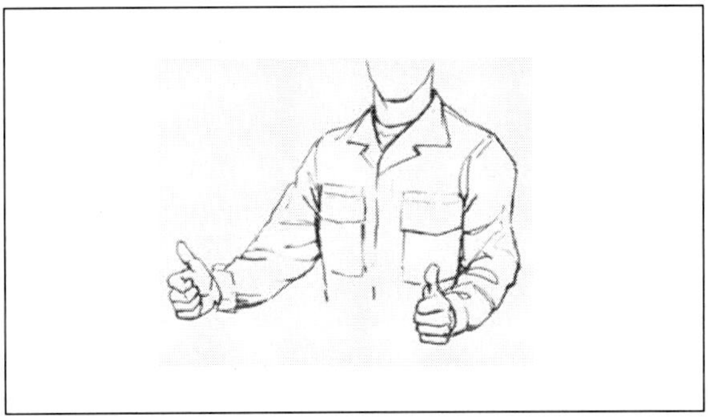

Figure 3-8. Arm-and-hand signal for GET READY.

(2) SIT IN THE DOOR (Figure 3-9). The rappel master extends both arms to the front with elbows locked, fingers extended, thumbs running along the index fingers, and palms facing downward. He bends slightly at the waist so that the arms are below the waist. Then, he moves both arms in a crisscross waving motion, alternating left over right and right over left.

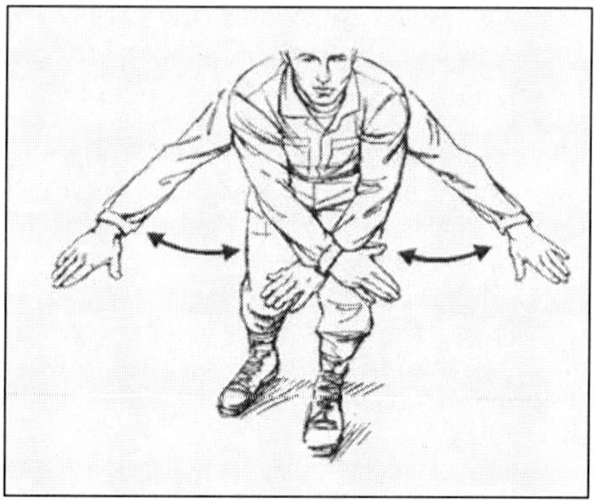

Figure 3-9. Arm-and-hand signal for SIT IN THE DOOR.

(3) THROW ROPES (Figure 3-10). The rappel master extends both arms to the front with elbows locked, fist clenched and index fingers extended. He points at respective rappellers, brings the forearms to an upright position and then down to the elbow-locked position.

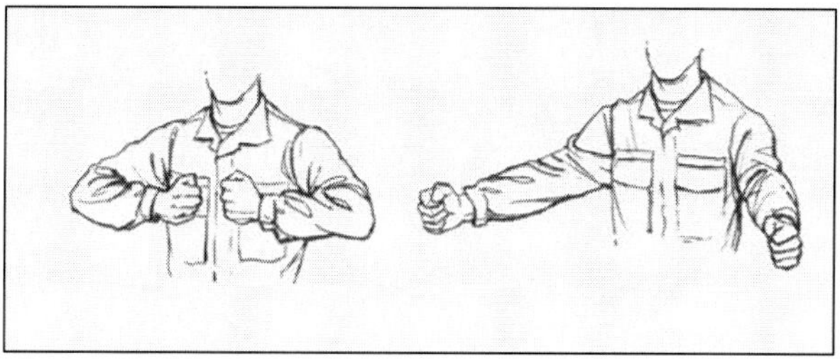

Figure 3-10. Arm-and-hand signal for THROW ROPES.

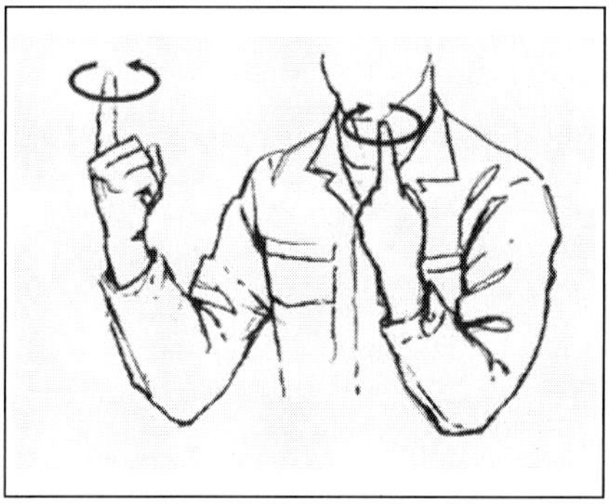

Figure 3-11. Primary arm-and-hand signal for POSITION.

(4) POSITION. The two arm-and-hand signals for POSITION are:

(a) Primary signal (Figure 3-11). The rappel master extends both arms to the front, elbows bent, forearms pointed upward, fists clinched, and index fingers pointing upward. He makes a circular motion with both forearms rotating in opposite directions.

(b) Alternate signal (Figure 3-12). The rappel master extends both arms to the front with elbows locked, fists clenched, and index fingers extended. He bends at the waist so that his arms are below his waist and makes a circular motion with arms rotating in opposite directions.

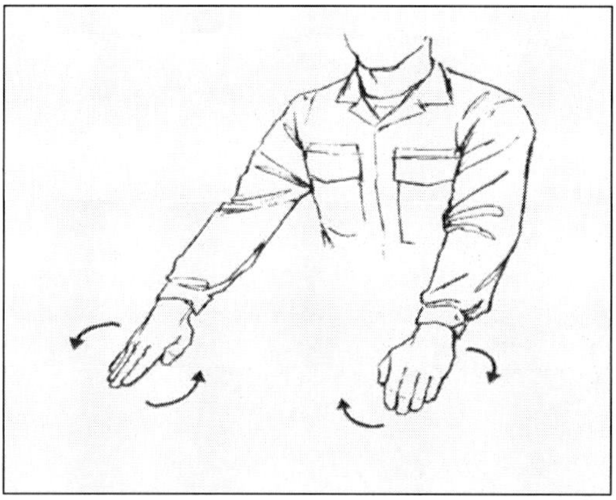

Figure 3-12. Alternate arm-and-hand signal for POSITION.

(5) GO (Figure 3-13). The rappel master extends an arm with elbow locked, fingers extended, and thumb along the index finger, and he points directly at the rappeller.

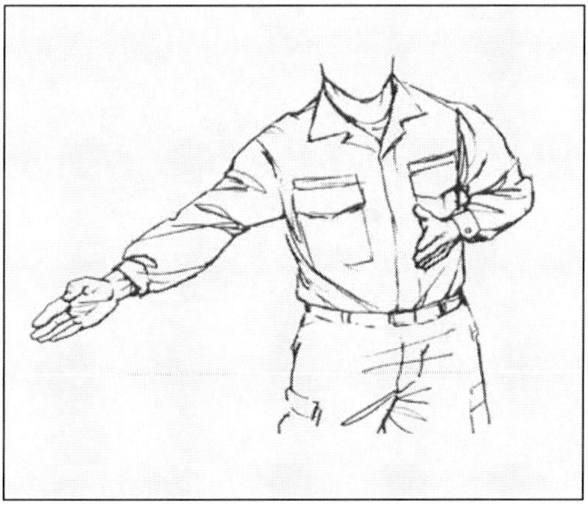

Figure 3-13. Arm-and-hand signal for GO.

3-21. INSPECTION AND SAFETY CONSIDERATIONS

The rappel master and pilot (or his representative) conduct a joint inspection of the aircraft to ensure the safety of all personnel and serviceability of equipment.

a. Cargo doors are locked in the open position or cleared for closing, depending on the mission.

b. All loose objects in the cargo compartment are removed or secured forward.

c. Sharp edges or protrusions on the cargo floor and door ledges are taped.

d. The donut ring and floating safety ring are serviceable and properly attached to tie-down fittings.

e. A headset and intercom jack for the rappel master are available and operational.

f. Serviceable safety harnesses are available for the rappel master and crew chief.

g. The recovery rope is installed and properly stowed.

h. Unused floor rings are taped down.

Section VI. RAPPELLING OPERATIONS FOR UH-60 BLACKHAWK HELICOPTER

The techniques used by the rappeller when rappelling from different aircraft are similar. However, positioning, seating, and the tie-down anchor point are different. Each rappeller is well trained on each aircraft before conducting a rappel.

3-22. CHARACTERISTICS

The UH-60 is a twin-engine, medium-speed, single-main rotor helicopter that can transport 10 rappellers, 1 rappel master, and 1 four-man crew when the center row of troop seats is removed. It is equipped with four 4,000-pound (load limit) cabin ceiling tie-down fittings or rings that are located in the ceiling of the troop/cargo compartment. These fittings are used as the primary rappelling rope anchoring points. The UH-60 is also equipped with eight 3,500-pound (load limit) cargo restraint net rings. Four of the eight rings are located in the ceiling of the troop/cargo compartment. These rings are used as the secondary rappelling rope anchoring points.

3-23. RIGGING OF UH-60 FOR RAPPELLING

The rappel master rigs the UH-60 helicopter by performing the following:

a. Locks both cargo doors in the open position.

Note: For arctic or other cold-weather operations, or during flights of long duration, the cargo doors are closed and locked until the time specified for opening them.

b. Removes the center row of troop seats.

c. Tapes any sharp edges or protrusions on the cargo floor and door ledges that may come in contact with the rappeller or the rappelling rope.

d. Stows loose equipment forward in the cargo compartment.

e. Extends the rappel master's intercom cord to the rear over the aft utility drain line and tapes the cord to the overhead troop seat support tube.

f. Installs the floor restraint provisions for rappellers No. 1 through 6.

g. Rigs and connects rappelling ropes to the aircraft's primary and secondary anchoring points.

(1) *Primary anchor points.* Ties a bowline with a half hitch about 4 feet from the standing end of the rope. Attaches the two primary snaplinks to the respective cabin tie-down fitting ring with gates facing in the opposite directions (Figures 3-14 and 3-16)

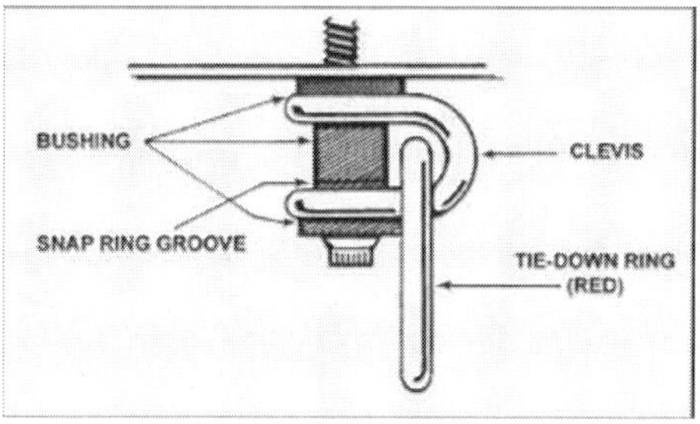

Figure 3-14. Primary snaplink attaching point.

(2) *Secondary anchor points.* Ties a bowline with a half hitch about 1 1/2 feet from the standing end of the rope. Attaches the secondary snaplink to the adjacent overhead cargo restraint net ring (Figures 3-15 and 3-16).

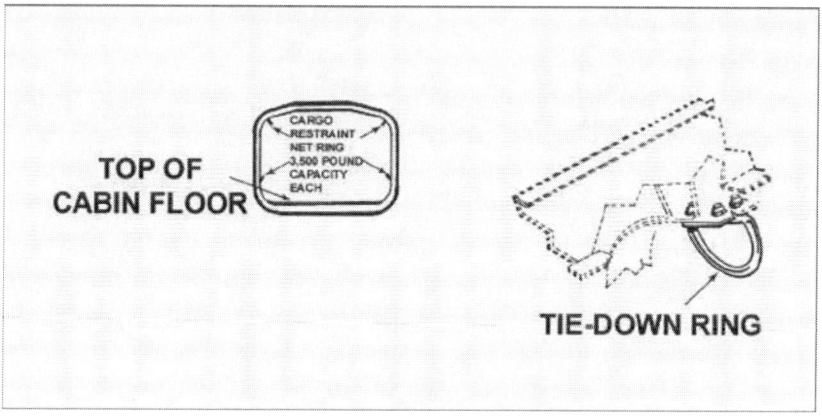

Figure 3-15. Secondary snaplink attaching point.

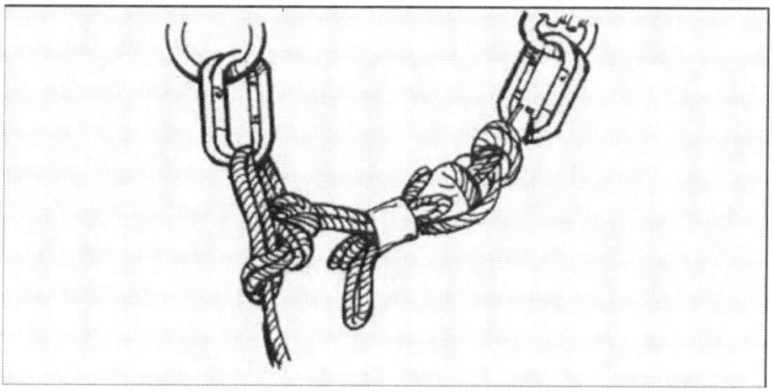

Figure 3-16. Primary and secondary snaplink attaching points.

h. Removes and secures the cargo hook access door and deploys the cargo hook in the DOWN position.

i. Installs the recovery rope for endangered rappellers.

j. Tapes the unused floor rings.

Note: There must be a minimum of 22 inches and a maximum of 30 inches between the anchor knots.

3-24. SEATING ARRANGEMENTS AND LOADING TECHNIQUES

Rappelling operations are executed quickly and safely to ensure the success of the operation in a combat or training environment. All rappellers must know the seating arrangements and the procedures for loading. They must be rehearsed under the supervision of the rappel master before conducting the operation.

a. A maximum of 10 rappellers, with and without combat equipment, and 1 static rappel master is seated and restrained aboard the UH-60 helicopter when the center row of troop seats are removed (Figure 3-17).

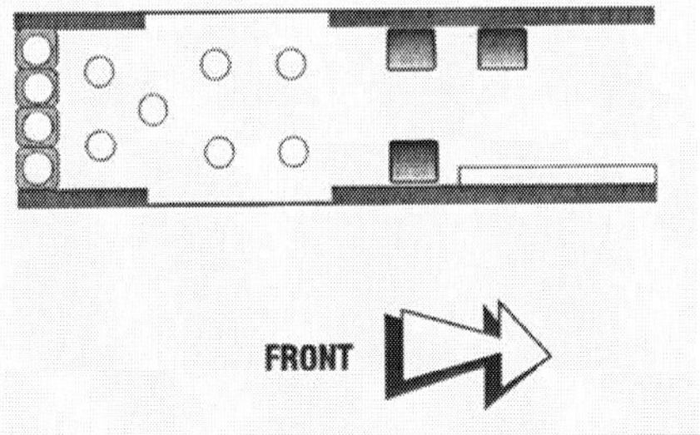

Figure 3-17. Rappelling personnel seating arrangement.

(1) Rappellers No. 1 through 6 and the rappel master are seated on the cargo floor. The rappellers are restrained using 3 1/2 feet of 1-inch tubular-nylon webbing (or equivalent) that is run through the cargo tie-down rings attached to the floor of the aircraft. The webbing is attached to the rappeller by means of a snaplink that is hooked to a Swiss seat. The rappel master is secured by his safety harness.

(2) Rappellers No. 7 through 10 are seated on the troop seats across the aft end of the cargo compartment. They are secured by the seat belts.

b. Rappellers are organized into 10-man sticks. They approach the aircraft from the left or right side in reverse order. They are divided into two groups and approach and enter the aircraft at the same time, depending on the situation. Rappellers No. 9, 7, 5, 3, and 1 enter the aircraft through the left door; they are seated and restrained. Rappellers No. 10, 8, 6, 4, and 2 enter the aircraft through the right door; they are seated and restrained. The static rappel master is restrained in the center of the cargo compartment using a C-3A troop-type safety belt or a safety harness during take-off and landing.

DANGER

BECAUSE THE LOWEST ARC OF THE ROTOR BLADE OCCURS AT THE DIRECT FRONT OF THE AIRCRAFT, APPROACHING THE AIRCRAFT AT THIS POINT COULD RESULT IN PERSONNEL INJURY OR DEATH.

3-25. RAPPELLING PROCEDURES

The rappel master ensures all personnel follow the rappelling procedures for the safe and efficient execution of the operation. These procedures are rehearsed under the supervision of the rappel master.

a. The rappel master is equipped with a headset and maintains communications with the pilot at all times. He wears a safety harness and stations himself in the center of the cargo floor to maintain control of the rappellers, their ropes, and all anchor attaching points.

Note: If all rappellers are rappelling into the same area, individual rappel ropes are not needed. The ropes used by rappellers No. 1 through 4 are used by the succeeding rappellers.

b. After all rappellers are in the aircraft, rappellers No. 1 through 4 hand the rappel master their primary and secondary anchor point snaplinks. The rappel master hooks the primary anchor point snaplink into the respective cabin tie-down fittings and the secondary anchor point snaplink into the respective cargo restraint net rings. After the primary and secondary anchor point snaplinks are secured, the first four rappellers hook onto their respective rappel rope. The rappel master checks the first four rappellers for the correct hookup.

c. After the first four rappellers exit and clear their ropes, the rappel master hands the next four rappellers their rappel ropes to hook up. Left-handed rappellers take extra precaution to ensure the correct hookup. Problems can be avoided by placing left-handed rappellers in the positions where they initially have the rope to their nonbrake side.

WARNING

Extreme care must be taken in hooking up and positioning personnel that require repositioning of their rope to their proper brake-hand side. This could result in bodily injury or damage to Army property.

d. After checking the rappellers for the correct hookup, the rappel master moves the next four rappellers into position on the floor of the aircraft. After the four rappellers exit and clear their ropes, the rappel master hands the last two rappellers their rappel ropes to hook up.

e. After checking the rappellers for correct hookup, the rappel master moves them into position on the floor of the aircraft. After the last rappeller is off the rope, the rappel master releases the ropes.

f. If the rappel master rappels, the crew chief is instructed to release or retrieve the ropes.

3-26. RAPPELLING COMMANDS

The UH-60 rappelling is conducted similar to the UH-1H rappelling with the exception that no skids are on the UH-60 on which to stand. Therefore, the edge of the floor, along the door of the helicopter, is used as a pivot point to assume the L-shaped rappelling position.

a. GET READY. This alerts the rappeller. The rappeller and rappel master perform final checks of the hookup, rappel seat, snaplink, and equipment.

b. THROW ROPES. The rappeller looks down below the aircraft to ensure that no one is under the aircraft. Keeping his brake hand in the small of his back, he tosses the deployment bag with the ropes *out and away* from the helicopter with his guide hand. The rappeller observes that the ropes are touching the ground and are not knotted or entangled.

c. POSITION. From a kneeling position with the brake hand in the small of the back, the rappeller rotates 90 degrees so that he is facing the inside of the aircraft and the rappel master. The rappeller then places his heels on the edge of the floor of the helicopter doorway and leans out into an L-shaped position. His feet are shoulder-width apart, the balls of his feet are on the edge of the helicopter doorway, his knees are locked, and his body is bent at the waist toward the helicopter.

d. GO. The rappel master initiates the rappel with this command. The rappeller flexes his knees and jumps backwards. At the same time, the rappeller throws his brake hand out at a 45-degree angle, letting the running ends of the ropes slide through both the brake hand and guide hand.

3-27. INSPECTION AND SAFETY CONSIDERATIONS

The rappel master and pilot (or his representative) conduct a joint inspection of the aircraft to ensure the safety of all personnel and serviceability of equipment.

a. Cargo doors are locked in the open position or cleared for closing, depending on the mission.

b. All loose objects in the cargo compartment are removed or secured forward.

c. Sharp edges or protrusions on the cargo floor and door ledges that may come in contact with the rappeller or his rappelling rope are taped.

Note: Do not tape the door latches or handles; this can interfere with door operations.

d. Primary and secondary rappelling anchor points are serviceable and securely attached to the aircraft structure.

e. A headset/helmet and intercom jack for the rappel master are available and operational, and the intercom extension cord is secured overhead.

f. Serviceable safety harnesses are available for the rappel master and crew chief.

g. The cabin ceiling tie-downs must have safety wire installed to ensure they do not come undone or unraveled, and the bolt head must be stamped H.

Section VII. RAPPELLING OPERATIONS FOR MH-53 HELICOPTER

The MH-53 helicopter is a highly versatile aircraft. It is used for a variety of missions, usually in conjunction with special operations.

3-28. CHARACTERISTICS

The MH-53 is a two-engine, single-rotor, heavy-lift helicopter. It has a crew of six and can be refueled while in flight. It has a precision navigational and communications package, which makes it excellent for conducting deep infiltration and exfiltration missions. The MH-53 helicopter can fly at night in all weather and terrain conditions. It can follow down to 100 feet and operate from unprepared sites. It is fitted with three gun stations for a mix of 7.62-mm miniguns and caliber .50 machine guns.

3-29. RIGGING OF MH-53 HELICOPTER FOR RAPPELLING

When rappelling from the crew entrance door, rappellers connect the ropes with locking snaplinks to the two 10,000-pound tie-down rings under the left scanner's window. When rappelling from the ramp, rappellers connect the ropes with locking snaplinks to the two 10,000-pound tie-down fittings on the left and right sides of the aircraft at station 522. Any sharp edges that could damage the ropes should be padded or taped. A length of fire hose may also be used over the portion of the rope that comes in contacts with the door or ramp edge.

3-30. SEATING ARRANGEMENTS AND LOADING TECHNIQUES

The MH-53 is a large, versatile helicopter. The seating arrangements and loading techniques are numerous. The using unit develops a seating and loading SOP that ensures the safety of all personnel and permits smooth, efficient execution of the mission.

3-31. RAPPELLING PROCEDURES

The rappel master or his assistant are at the two exit points (ramp and door). The rappel master maintains communications with the helicopter commander and relays all commands and time warnings. The commander issues time warnings at 20-minute, 10-minute, 5-minute, and 1-minute intervals. During limited visibility, the rappel master may use NVDs to observe the safety/belayer. Chemical lights are attached to the top or bottom of the rappel rope. Arm-and-hand signals should be IAW The Air Assault School Handbook and FC 57-2.

Note: Night vision goggles cause limited depth perception and a tunnel vision effect.

a. The rappel master is secured by a harness and ensures the proper hook up of each rappeller. Once hooked up, the rappellers release their safety belts. At the command of the rappel master, the rappellers position themselves to ease immediate deployment.

WARNING

Before conducting deployments off the ramp, the deploying team members must be briefed on the importance of maintaining separation between members. About 24 to 27 inches are desired. This separation helps maintain aircraft center of gravity limits.

b. The rappel master deploys ropes only after the helicopter is in a stable hover over the target area and he has given the command ROPES. He ensures the ropes are on the ground before giving the command GO. After the last rappeller is off the rope, the rappel master retrieves or cuts the ropes away.

3-32. RAPPELLING COMMANDS

The helicopter rappelling commands for the MH-53 are as follows:

a. GET READY. This command is given as the helicopter approaches the rappel site. It alerts the rappellers of the approach to the site and signals them for a final equipment check.

b. POSITION. This command clears the first rappellers into position for deployment.

c. THROW ROPE. This command is given once the helicopter reaches a stable hover.

d. GO. The rappellers exit.

3-33. INSPECTION AND SAFETY CONSIDERATIONS

The rappel master briefs all personnel participating in the operation. He conducts hands-on inspection to ensure the safety of all personnel and serviceability of equipment.

a. Tie-down fittings are serviceable.

b. All sharp or protruding edges that may come in contact with the rappelling ropes are padded or taped.

c. All ropes are retrieved or cut away before forward movement of the helicopter.

d. Only three rappellers are deployed at a time—two from the ramp and one from the personnel door.

CHAPTER 4
SPECIAL PATROL INFILTRATION/EXFILTRATION SYSTEM

The special patrol infiltration/exfiltration system (SPIES) provides an excellent form of exfiltration over short distances. It is not recommended for infiltration because team members are exposed the entire time. Due to the nature of SPIES operations, a thorough briefing is required for all participants before the operation. Personnel being extracted must train extensively in the SPIES extraction before infiltration. For the other personnel involved, a complete preoperations briefing is held before the operation. This is crucial when additional assets are involved other than the extraction helicopter such as gunships, aerial observers, or artillery support.

Section I. SPIES MASTER DUTIES AND QUALIFICATIONS

The SPIES master holds the rank of sergeant or above (may be waived by the commanding officer). The commanding officer ensures that this qualification is entered on the soldier's training record.

4-1. QUALIFICATIONS

To qualify to be a SPIES master, the sergeant must participate in at least three SPIES operations (observe twice and execute SPIES master duties once under observation). For example, he hooks up the helicopter, assists in the preparation of an operation, and conducts a successful operation under the supervision of a qualified SPIES master. A SPIES master is conscientious and thorough, and he knows all aspects of a SPIES operation. He can give an effective pilot's brief, use the aircraft communications equipment, and understand aviation terminology.

4-2. DUTIES

The SPIES master is responsible for the safe conduct of the SPIES operation.

a. **Preflight Duties.**

(1) Inventories and inspects all SPIES equipment.

(2) Briefs the pilot and other concerned personnel about details of the operation, especially the extraction and dismounting procedures.

(3) Ensures that an ICS helmet and gunner's belt are available for the SPIES master's use. Uses a sling rope if no belt is available. Connects and checks the operation of the ICS to be used. (ICS communications is established between the SPIES master and pilots on all SPIES operations.)

(4) Attaches the SPIES rope to the helicopter IAW the guidance in this chapter.

(5) Ensures that nothing is adrift in the aircraft that may fall on a team member.

(6) Checks the location of the emergency axe. Places the axe where readily available, yet secure enough so as not to endanger the men on the SPIES rope. (Inspects the axe to ensure that it is sharp.)

b. **Extraction Duties.**

(1) On arrival at the team's estimated position, assists the pilot in determining the exact location of the team members.

(2) As the aircraft approaches the team's location, aids the pilot (using the clock system) in placing the aircraft directly above the team.

(3) Requests permission from the pilot to drop the SPIES rope when the aircraft is hovering above the team.

(4) Drops the rope, taking care to avoid striking team members on the ground.

(5) Notifies the pilot when the rope is down, and reports all altitude corrections to ensure that team members reach all SPIES attachment points.

(6) Watches for the "thumbs-up" signal from all team members.

(7) On receipt of the "thumbs-up" signal, advises the pilot that the team is ready for extraction and requests a vertical lift-off.

(8) Advises the pilot of the team's position, the location of any potential obstacles, and the avoidance of horizontal movement.

(9) If a team member becomes entangled with an obstacle during the extraction, notifies the pilot and requests that the vertical lift be stopped. If the situation is critical, prepares to cut the SPIES rope (the anchor point or cargo straps) after the team members are secured to the obstacle or on the ground.

(10) When positive that all obstructions are clear, advises the pilot to obtain a safe altitude (about 350 feet AGL for training purposes or as the situation dictates in combat) or to transition into forward flight.

(11) At frequent intervals during the flight, advises the pilot on the safety status of all team members. Maintains a constant visual watch on the team and checks security of the SPIES attachments often.

c. **Dismounting Duties.**

(1) On arrival at the dismounting area, informs the pilot as to the approximate distance of the lower rope end from the ground.

(2) Once the pilot starts the vertical descent, continually informs him as to the approximate distance of the lower rope end from the ground.

(3) Informs the pilot of any horizontal drift that occurs and any obstructions near the SPIES rope. Also informs the pilot of any oscillation that may occur.

(4) Informs the pilot when the rope is about 25 feet above the ground and again when it is 10 feet above the ground. Ensures that the rate of descent is slow enough to enable the team members to land and get out from under team members safely.

(5) Reports when the first man initially touches down, when the last team member starts to safely move away from under the helicopter, and when all team members are disconnected.

(6) On order of the pilot, either retrieves the SPIES rope into the helicopter or disconnects the SPIES rope and drops it to the ground. When using the UH-1H/UH-60 helicopter, the only way to retrieve the SPIES rope while in the air is by having an arranged recovery rope attached with a 16-foot sling rope. In some cases, the SPIES master joins two 12-foot-long sling ropes to haul the SPIES rope aboard and attaches the rope about 5 or 6 feet below the cargo hook or cargo strap hookup point. The type of knot used to connect the sling (or recovery) rope to the SPIES rope is self-tightening in nature (for example, the prusik knot). The SPIES master fastens the standing end of the sling rope to the deck tie-down or uses a snaplink. Although it is important to keep the line out of the way, the primary consideration is its length. The rope must be long enough for any oscillation in the SPIES during flight.

Section II. PREOPERATIONS BRIEFINGS AND PROCEDURES

Before conducting a SPIES training mission, the participants must have a basic understanding of the requirements. The SPIES master conducts briefings to ensure the soldier extracted and the pilot know the procedures. As SOPs are developed and units train together, the SPIES master simply refers to the SOP. He always gives a safety briefing and conducts an equipment inspection.

Note: SPIES has not been approved for Army-wide use. Approval must be obtained from division level or higher commander before the conduct of any training.

4-3. SAFETY

As in all training, a safety briefing must precede operations using the SPIES. The briefing should consist of, but not be restricted to, a review of the following:

- Characteristics of equipment associated with the SPIES.
- Equipment inspection.
- Proper donning of the harness.
- Methods of extraction and insertion used.
- Emergency signals.

4-4. COMMUNICATIONS

Due to the noise level associated in all helicopter operations, radios are used for communication. They are also used before the arrival of the helicopter. Precise arm-and-hand signals are established in case of radio failure or poor communications due to static or noise overriding the audio output of the radio. During the first part of the operations, the SPIES master either observes (daytime) or knows that a definite procedure is taking place (night or jungle) while the teams are hooking up to the SPIES rope for extraction. Headsets and voice suppressors replace the handset for better radio procedures. This allows the RATELO on the ground to use both hands while the helicopter hovers for a faster and safer hookup. If radio communications are hampered, special procedures are used along with hand or light signals.

4-5. EXTRACTION PROCEDURES

After the team has been located, the SPIES master assists the pilot in directing the helicopter to the proper distance over the team.

a. The team leader positions himself so that he can move and approach the rope as it is dropped by the SPIES master. Once the rope is clear of any obstacles, the team leader signals the team to their assigned positions along the 10 hookup points. Using the primary or harness snaplink, each team member hooks to the D-ring on his side of the line. This is the primary hookup. Once the team is hooked up, the team leader hooks into the alternate or second hookup point using the safety line and snaplink. He then faces forward along the line so that he is heading in the direction he is traveling when the aircraft starts its ascent.

b. The SPIES rope is held up and routed over the shoulder closest to the rope. With the other hand, the team members give thumbs-up signals to allow both the team leader and the SPIES master to see that they are ready to go. The team leader physically inspects (if time and situation permits) or hooks himself in on the lowest point along with the RATELO to ensure the running end is clear of all obstacles. He gives the thumbs-up signal to the SPIES master. This thumbs-up signal (at night, an arranged light signal) continues until the team is lifted off the extraction surface. The team leader also signals (arms held straight to the sides) when the team is at safe altitude (approximate 10 feet above the tallest obstacle at the extraction site). The helicopter starts a transition in a horizontal direction on its return flight.

4-6. EMERGENCY PROCEDURES

During the flight, from extraction until the team is safely and quickly detached from the SPIES rope, each team member should be aware of any problems that arise from above or below. The man above checks the man below. At the first sign of danger or if an emergency arises, the team leader or a team member places his free hand on his head. Upon observing anyone on the SPIES rope with his hand on his head, the SPIES master tells the pilot to make an emergency landing in the nearest and safest area.

4-7. DISMOUNTING PROCEDURES

During the familiarization training phase, the SPIES master ensures all members know they should immediately head in the direction of the nose (12 o'clock) of the aircraft. This allows the pilot to see that the team is out from under the aircraft. When an emergency situation arises, the pilot can make a better appraisal of the situation if he can see all the members of the team. When the helicopter is making a scheduled landing, the team ensures that the SPIES rope does not interfere with the aircraft and that the aircraft does not land on the rope.

4-8. INSPECTION OF EQUIPMENT

The SPIES is inspected by a certified SPIES master at six-month intervals and whenever the SPIES master questions serviceability. Outdated, spliced, abraded, or cut rope is removed from service. However, loose or broken stitching in excess of three stitches is not repaired.

> **CAUTION**
>
> **Acid contamination, cuts, or fraying of harness or sling webbing constitute nonrepairable damage.**

a. The certified SPIES master inspects the harness and suspension sling webbing for signs of contamination from oil, grease, acid, or rust. He also checks for cuts, twists, fading, excessive wear, fusing (indicated by unusual hardening or softening of webbing fibers), fraying, burns, abrasions, and loose or broken stitching. The damaged harness or suspension sling is removed from service and returned to supply for appropriate disposition.

b. The SPIES master inspects all hardware for signs of corrosion, pitting, ease of operation, security of attachment, bends, dents, nicks, burrs, and sharp edges.

Replacement of hardware (except chest strap adapter) that requires unstitching of webbing renders the harness unserviceable.

c. The SPIES master replaces the V-ring by cutting the strap above the stitching. He folds and stitches a new end section of the leg strap. If the leg strap is damaged, the SPIES master returns the harness or suspension sling to supply for appropriate disposition.

d. The SPIES master checks ropes, harnesses, and suspension slings for expiration of service or total life, which is 7 years of service (opening manufacturer's package) or 15 years from date of manufacture, whichever occurs first. The SPIES master also ensures the rope is free of splices. He ensures the eye loop at the end of the SPIES rope is not broken, frayed, or loose.

e. The SPIES master inspects the rope surface for cuts, excessive abrasions, and snags. Cuts on the rope are excessive when there are four or more cut strands in any 5-inch length. The two-to-one braided rope has 12 pairs of strands (24 individual strands) around the circumference. Abrasion is extensive when torn yarns are equivalent to that of four strands of any 5-inch length. Rope that has been subjected to heavy loads may display glazed areas where it has worked against hard surfaces. This condition may be caused by paint or fused fibers. After long use, the rope may become fuzzy on the surface (although this should be minimized with the surface coating). However, the effect on the rope's strength is negligible.

f. The SPIES master inspects the rope for signs of contamination by acid, alkaline compounds, salt water, fire extinguishing solutions, and petroleum-based solvents. Although the ropes gradually change color, such changes do not indicate decrease in strength unless the change is due to contact with strong chemicals. Changes in color caused by chemicals are usually spotty; changes caused by use are uniform throughout the length of the rope.

4-9. RIGGING OF UH-1H HELICOPTER FOR SPIES OPERATION

Personnel who are not familiar with SPIES are encouraged to observe or take part in the rigging of the helicopter. This builds confidence in the equipment and assists in a more comprehensive training of new SPIES masters.

a. **Equipment.**

- One 120-foot SPIES rope with deployment bag.
- Two 11-foot, 3- or 4-loop cargo slings or two 9-foot, 3- or 4-loop cargo slings.
- Two Type IV connector links (four if aircraft does not have cargo hooks).
- Heavy duty tape (100-mph tape).
- One 12-foot sling rope.
- Five oval snaplinks (nine if the aircraft does not have cargo hooks).

NOTE: If the aircraft does not have cargo hooks, use four 11-foot or four 9-foot, 3- or 4-loop cargo slings.

b. **Installation.**

(1) The primary attachment point for the SPIES rope is the cargo hook. The end of the SPIES rope having a polyurethane encapsulated eye is attached to the cargo hook.

The two 9- or 11-foot long cargo suspension slings are joined to form one continuous sling, using a Type IV link. The sling is stretched out on the helicopter deck. One end is taken under the helicopter and through the eye of the SPIES rope. It is then connected on the other end of the sling using a Type IV link assembly. The sling must pass between the helicopter skids and the fuselage.

(2) Once the the SPIES rope and cargo straps are in place, the straps running across the deck of the helicopter are secured in place by at least four and as many as eight snaplinks. The snaplinks are evenly spaced across the deck and alternated from one side of the strap to the other and top and bottom, so that the first snaplink is to the rear of the strap and going around the bottom two straps. The next snaplink is in the front of the cargo strap and around the top two sections of the strap. This process continues until at least four points are established.

(3) If eight snaplinks are available, then each tie-down has two reversed. If no hook is available or is not working properly, the SPIES can be used safely by doubling the cargo slings and Type IV links. Thus, two cargo straps are side by side for a total of four slings and four Type IV links. Padding is used around the edge of the fuselage to protect the sling from damage.

(4) Care must be taken when using the UH-1H helicopter because of the ways in which it can be outfitted. Some helicopters have a step attached at the doorway. This is an added obstruction not only during installation but also during the operation. Others may have rocket pods or machine guns mounted. Not all of the UH-1 helicopters are hooked up the same way each time.

4-10. RIGGING OF UH-60 HELICOPTER FOR SPIES OPERATION

The SPIES master inspects all materials required to rig the helicopter.

a. **Equipment.**

- One 120-foot SPIES rope with deployment bag.
- Two 11-foot, 3- or 4-loop cargo slings or two 9-foot, 3- or 4-loop cargo slings.
- Two Type IV connector links (four if aircraft does not have cargo hooks).
- Heavy duty tape (100-mph tape).
- One 12-foot sling rope.
- Five oval snaplinks (nine if the aircraft does not have cargo hooks).

Note: If the aircraft does not have cargo hooks, use four 11-foot or four 9-foot, 3- or 4-loop cargo slings.

b. **Installation.**

(1) The primary attachment point for the SPIES rope is the cargo hook. The end of the SPIES rope has a polyurethane encapsulated eye that is attached to the cargo hook. The two 9- or 11-foot cargo suspension slings are joined to form one continuous sling using a Type IV link. The sling is stretched out on the helicopter deck. One end is taken under the helicopter and through the eye of the SPIES rope. It is then connected on the other end of the sling using a Type IV link assembly. The sling passes between the helicopter skids and the fuselage. Padding is used around the edge of the cargo hatch to protect the sling from damage.

(2) Once the SPIES rope and cargo straps are in place, the straps running across the deck of the helicopter are secured in place by at least four and as many as eight snaplinks. The snaplinks are spaced evenly across the deck and alternated from one side of the strap to the other and top and bottom. Thus, the first snaplink can be to the rear of the strap, wrapping around the bottom two straps. The next snaplink is in the front of the cargo strap and around the top two sections of the strap. This process continues until at least four points are established.

(3) If eight snaplinks are available, each tie-down has two snaplinks connecting the same spot, and the swing gates are reversed. If no hook is available or is not working properly, the SPIES can be used safely by doubling the cargo slings and Type IV links. Two cargo straps are side by side with a total of four slings and four Type IV links.

4-11. RIGGING OF CH-46/CH-47 HELICOPTER FOR SPIES OPERATION

The SPIES master inspects all the materials required to rig the helicopter.

a. **Equipment.**

- Two 11-foot, 3-loop slings.
- Two 9-foot, 3-loop cargo slings.
- Four Type IV connectors.
- One 13-foot sling rope.

b. **Installation.**

(1) The SPIES rope is attached using two 9- or 11-foot cargo suspension slings and four Type IV links. The cargo slings are passed through the encapsulated eye of the SPIES rope and attached to the outboard cargo tie-down rings on the aircraft floor. Two tie-down rings are used for each sling. Padding is used around the edge of the cargo hatch to protect slings from damage.

(2) Not all of the tie-down rings will be in the exact same position on the helicopters. This is one of the main considerations in deviating from the installation procedures. When possible, the cargo straps are placed to form two U shapes. One strap is placed forward of the cargo hole in the center of the aircraft floor and one aft or toward the rear of the helicopter. Installed properly, the cargo straps hold the SPIES rope comfortably in the center of and slightly below the opening of the cargo hatch.

(3) The use of snaplinks attached close to all four faulty tie-down points not only ensures a backup in case of a faulty tie-down ring but also reduces the amount of movement in the cargo suspension straps. A total of eight snaplinks are used for added security at each point with the swing gates reversed.

Section III. LAND AND WATER EXTRACTION

If the situation, mission, or terrain suggests the possibility of a SPIES extraction, the leader includes the SPIES harness in each individual's equipment list. If the mission or insertion precludes the wearing of the harness, it will be carried on the top and just inside the flap of the pack being used. Once the extraction helicopter has been requested, the harness is retrieved and donned.

4-12. LAND EXTRACTION PROCEDURES

The SPIES is used only when the team requires immediate extraction or is unable to move to a clear (open) position suitable for helicopter landing.

a. The extraction helicopter proceeds to the area and establishes radio or visual contact with the team. The backup helicopter equipped with the SPIES remains aloft and away from the area. The helicopter maintains visual contact with the LZ and monitors radio communications.

b. After the team has indicated readiness for pickup and the tactical situation has stabilized, the extraction helicopter moves to the LZ by the safest route. When the helicopter is above the team's location, the SPIES master drops the SPIES rope on order of the pilot after the aircraft has obtained a stable hover at slightly above treetop height.

c. The team hooks up the same as in the familiarization procedures. Team members sling individual weapons over the shoulders, securing weapons and equipment to withstand the wind. A safety line is attached to prevent lost weapons during SPIES operations. When the team is ready, the team leader gives a thumbs-up signal.

d. During extraction, the team RATELO maintains communications with the extraction helicopter. He gives a verbal backup to the thumbs-up signal and relays all information during the flight. His location is near or at the bottom hookup point so he can assist in giving accurate information about the extraction, clearing of obstacles, and descent.

e. Lift-off of the extraction helicopter is vertical until the SPIES rope has cleared all obstacles. With the barrel directed downward at a 45-degree angle and outward, the team members fire their individual weapons using the hip position.

f. Once the SPIES rope is clear of all obstacles, the extraction helicopter changes to horizontal flight and departs the area by the safest route. Airspeed is limited to 70 knots in moderate climates and 50 knots in cold climates while team members are attached to the SPIES rope.

g. When the extraction helicopter has reached a safe dismount area, a transaction is made into hover flight at an altitude of 250 feet AGL. A vertical descent is started with the SPIES master continuously providing information to the pilot on the distance from the ground to the lower end of the SPIES rope. The vertical descent rate of the aircraft (at touchdown) is less than 5 feet per second.

h. When the team reaches the ground, they immediately move out toward the front of the helicopter. All team members rapidly unhook themselves and teammates that need assistance. Once this is completed, they either move away from the area and set up security or assist in clearing the rope if the helicopter is going to land.

4-13. WATER EXTRACTION PROCEDURES

The SPIES is suitable for extracting teams from the water. For this procedure, three inflatable life vests or any type of floatation device is tied to the SPIES rope to provide buoyancy for the rope while in the water. One floatation device is tied at each end of the attachment points, and one floatation device is tied in the middle of the attachment point area just above the middle two sets of D-rings. The team members to be extracted wear their SPIES harnesses under their individual life vest. They can also wear swimming fins, masks, and snorkels (amphibious operations) to ease hooking up to the SPIES rope within the spray area beneath the hovering helicopter.

a. After the extraction helicopter has attained a stable hover above the team member's location, the SPIES master drops the SPIES rope (with floatation attached) on order from the pilot. When the team members complete the hookup to the SPIES rope, the team leader signals the SPIES master to start lift-off. Lift-off of the extraction helicopter is vertical until all team members and the bottom end of the rope have cleared the water. During the initial lift-off, team members know that they are to be dragged through the water. The team members prepare to roll on their backs until clear of the water.

b. Flight speed and altitude are the same as over land. The dismounting procedures remain the same, except when landing on a ship. Once on board, the team members take their orders from personnel in charge of the deck.

Section IV. AFTER-OPERATIONS PROCEDURES

After-operations procedures include repairing, cleaning, and proper storage of the rappelling equipment.

4-14. REPAIRING AND CLEANING OF EQUIPMENT

The contaminated ropes are washed with a mild detergent, such as liquid dish soap, and cold water followed by a rinse in clean, fresh water. Ropes are dried at a temperature not to exceed 140 degrees Fahrenheit. Stubborn oil, grease, hydraulic fluid, and other petroleum stains can be removed with the cleaning agent xylene (Grade A or B, TT-X 916).

4-15. STORAGE OF EQUIPMENT

To avoid ultraviolet deterioration, the nylon materials should be protected from direct sunlight. The SPIES rope is stowed in an aviator's kit bag for protection when not in use. Bins or similar facilities are used for storage of SPIES equipment. Shelves used for storage should be at least 4 inches from the walls and 12 inches from the floor. Areas used for storage should be well ventilated and free of oil, acid, cleaning compounds, and other contaminants. Equipment must *not* be stowed above or near hot water pipes or heating apparatus.

CHAPTER 5

FAST-ROPE INSERTION AND EXTRACTION SYSTEM

The fast-rope insertion and extraction system (FRIES) is employed by small units to make rapid infiltration and exfiltration using rotary-wing aircraft in confined areas. The use of FRIES is normally restricted to situations where aircraft cannot land, and when mission requirements preclude the need for large amounts of equipment, and ammunition or heavy crew-served weapons.

Note: Headquarters, Department of the Army (HQDA) policy specifies that FRIES is not approved for Army-wide use and names the Commanding General, United States Army Special Operations Command (CG USASOC) as the executive agent for FRIES doctrine. The use of FRIES is restricted to special operations forces, pathfinders, long-range surveillance units, and HQDA-approved schools with a USASOC-approved FRIES program of instruction. Requests for other units to conduct FRIES operations should be submitted to Commander, USASOC, ATTN: AOOP-TRS, Fort Bragg, NC 28307-5000.

Section I. GENERAL

FRIES information is subject to change. Units that recognize the need to update this chapter should submit recommendations to Commander, USASOC, ATTN: AOOP-TRS, Fort Bragg, NC 28307-5000.

5-1. OBJECTIVES

The objectives of this chapter are to—

- Prescribe the requirements for FRIES qualification and training, including proficiency sustainment, safety, and administrative procedures.
- Establish the scope of the duties and responsibilities of key personnel who conduct FRIES operations.

5-2. GUIDANCE FOR COMMANDERS

Units that have HQDA approval to perform FRIES operations are authorized to conduct initial FRIES qualification and FRIES master (FRM) qualification training. Training requirements are established in this publication, USASOC Regulation 350-6, and applicable SOAR policies. Aircraft material requirements are specified by the US Army Technology Applications Program Office (TAPO) .

a. FRIES training and operations are inherently risky; therefore, safety is paramount. All training and operations require in-depth attention to detail from planning through preparation and execution.

b. The following basic guidance applies to FRIES training and operations:

(1) Unit commanders must personally approve FRIES training sites.

(2) A risk analysis and or assessment must be completed before FRIES operations.

(3) Night FRIES operations are medium risk or higher.

(4) Units authorized to conduct FRIES operations may incorporate FRIES activities in field training exercises only after personnel have become FRIES qualified.

(5) Field exercises using FRIES activities are subject to the limitations imposed by applicable directives, including mandatory support requirements.

(6) Requests for waiver of procedures or restrictions prescribed herein must be submitted in writing with endorsement of the first general officer in the requesting unit=s chain of command to HQ USASOC.

5-3. TRAINING PREREQUISITES

CAUTION
Trainers must ensure all personnel preparing to participate in FRIES training are tested *with equipment* to ensure they have sufficient upper body strength to safely perform the full scope of roping duties.

Before allowing soldiers to participate in FRIES training and operations, the unit commander must make sure the soldiers are physically fit. Minimum screening standards that apply to each training participant include—

- Having passed the Army physical fitness test within the past six months.
- Having passed a medical examination within the past two years.
- Being free of any injury or physical condition that could cause a potential safety hazard during FRIES training.
- Demonstration of ability to perform a controlled descent from a height of 15 feet while carrying a 40-pound load (not including the basic duty uniform and combat boots).
- Demonstration of ability to hold a static position on a FRIES rope for 20 seconds, using hands and feet to lock in while carrying a 40-pound load.
- Being qualified and current to fly FRIES activities IAW HQ USASOC and the Special Operations Aviation Regiment (SOAR) Aircrew Training Program and applicable policies (for aircrews).

Note: The peacetime maximum soldier load will not exceed 50 pounds. This includes helmet, weapon, vest, web gear, and rucksack. Rucksack weight will not exceed 35 pounds.

5-4. PERSONAL EQUIPMENT REQUIRED

The minimum personal equipment required for FRIES training or operations includes—

- Heavy leather gloves.

- A helmet with a chin strap.
- Protective goggles.
- A long-sleeve shirt or jacket, long pants, and boots.
- Hearing protection and identification tags for helicopter operations.

Section II. FRIES QUALIFICATION TRAINING

Each major unit (commanded by an O5 or higher) is responsible for conducting its own FRIES qualification and sustainment program. FRIES trainers will be FRIES master (FRM) qualified and current. Unit training should be tailored to unit and situation needs, however proponent requisites will not be reduced without written approval of the CG USASOC.

5-5. INITIAL FRIES QUALIFICATION TRAINING

Before participating in fast-rope operations, personnel are briefed on the FRIES and its purpose, capabilities, and limitations. The briefing also covers the duties and responsibilities of the pilot in command (PIC), safety, FRIES master (FRM), the assistant FRIES master (AFRM), and any ground assistants. Once the FRIES briefing is conducted, the remainder of the initial training is hands-on practice of the proper FRIES operational techniques.

a. Individuals are shown the proper techniques for boarding the aircraft, moving to the door, grasping and descending, locking-in, and clearing the rope. Individuals are also shown the proper wear, hook-up, and use of the STABO extraction harness.

b. After the demonstration, all ropers participate in a practice exercise in which they properly perform the following tasks using a tower and subsequently an aircraft. All ropers must demonstrate a minimum of six properly executed FRIES descents (three without equipment and three with equipment) and at least one successful lock-in.

(1) *Perform FRIES descent and lock-in from a tower at the 34-foot level.* Training should be progressive, starting from a tower at the 34-foot level without equipment and then with equipment. All ropers must complete a successful lock-in at this level. Ropers will not progress above the 34-foot level until after demonstrating the ability to stop descent, lock-in, and hold a stationary position for 20 seconds with equipment.

(2) *Perform FRIES descent and lock-in from a tower at the 60-foot level.* The recommended succession will be no less than three successful FRIES descents on the 60-foot tower, again starting without equipment, completing a lock-in, and finishing with equipment.

(3) *Perform FRIES descent from a helicopter at 60 feet or higher.* The recommended completion should be at least two helicopter descents from 60 feet or higher (one without equipment and one with equipment).

WARNING

During initial qualification, students will not perform rapid exits or have more than three soldiers on a rope at any time. During advanced training, the limit is no more than five personnel with equipment on any rope at any time. In all cases (insertion or extraction), the maximum load allowed on any fast-rope system will not exceed 1,500 pounds.

CAUTION

All ropers should be limited to no more than a total of 10 roping events in a 24-hour period; no more than 6 events should be with equipment.

5-6. FRIES PROFICIENCY SUSTAINMENT TRAINING

Commanders must make sure soldiers participating in FRIES operations receive sustainment training on equipment and procedures within 24 hours before the FRIES operation. Ropers who do not attend FRIES sustainment training will not be allowed to participate in FRIES operations. As a minimum, training will include a review of the following:

- Arm-and-hand signals.
- Individual equipment riggings.
- Aircraft familiarization.
- Safety procedures.
- Any rehearsals the FRM or commander deems necessary.

Section III. FRIES MASTER SELECTION AND QUALIFICATION TRAINING

Selection and qualification of FRIES masters (FRMs) is a unit prerogative. Units will maintain records of qualification and proficiency. These records will accompany SM to new units when a permanent change of station occurs.

5-7. FRIES MASTER SELECTION

Units may train and certify soldiers as FRIES-qualified FRMs. Soldiers selected by units to qualify as FRMs should be the rank of E5 or higher and should have demonstrated leadership abilities with emphasis on maturity and decisiveness.

5-8. FRIES MASTER PREREQUISITES

Soldiers performing FRM, AFRM, and safety duties must have completed initial FRIES qualification training, be experienced in FRIES operations, have completed FRM training, and have executed at least 10 FRIES descents and 2 extractions. An FRM candidate's experience must include no less than five descents with combat load and at least one night descent with equipment from 60 feet or higher.

5-9. FRIES MASTER TRAINING AND CERTIFICATION

During FRM training and certification, FRM candidates must participate in three FRIES helicopter operations (twice observing another FRM and once executing FRM duties under observation of a current FRM) and be certified by either the parent unit or an authorized school. FRMs must be proficient in FRIES operations and must demonstrate proficiency at the following:

- Inspecting, preparing, and rigging all FRIES gear.
- Inspecting and preparing aircraft for FRIES operations.
- Controlling and coordinating actions of AFRMs, safeties, and ropers.
- Preparing and conducting both insertion and extraction operations.
- Conducting troop and pilot briefings.
- Giving arm-and-hand signals.
- Giving time warnings and associated commands.
- Deploying and retrieving FRIES ropes.
- Rigging and lowering equipment.

5-10. FRIES MASTER REFRESHER TRAINING

Units will conduct refresher training to maintain acquired skills. FRMs who have not participated in FRIES operations during the past six months will receive refresher training by a current FRM and serve as an AFRM before performing FRM duties. Refresher training for ropers consists of an FRM briefing and participation in FRIES training.

Section IV. KEY PERSONNEL DUTIES AND RESPONSIBILITIES

The following personnel duties and responsibilities provide baseline requirements for the safe conduct of FRIES operations. Unit standing operating procedures (SOPs) may increase (but will not reduce) training safety requirements.

5-11. AIR MISSION COMMANDER

The air mission commander (AMC) provides overall command and control when conducting large, complex missions. The AMC is selected or appointed by the headquarters responsible for the overall mission. He ensures that all operators, aircrew members, and support elements synchronize their actions during the conduct of the mission and the FRIES operation. An AMC is not usually needed during small-unit qualification or proficiency training events when only one helicopter is used.

5-12. FRIES TRAINING OFFICER

Units that conduct FRIES training appoint a training officer in charge (OIC) or a noncommissioned officer in charge (NCOIC) to manage the training. The OIC is responsible for planning and coordinating various aspects of training to include—

- Coordinating all support activities, such as procuring FRIES equipment, aircraft, training areas, medical support, and communication.

- Assigning qualified personnel to perform FRM, AFRM, ground safety, and medic duties.
- Adhering to procedures for planning, preparing, and executing the operation IAW this TC, USASOC Regulation 350-6, and the participating unit=s SOPs.
- Ensuring the FRMs, AFRMs, PICs, aircrews, safeties, and medics are briefed on the operation.
- Ensuring FRIES training events are not conducted on or over unsafe ground. (The ground must be free of loose material that could become airborne and cause an unsafe condition or obscure vision for aviators or ropers.)
- Ensuring a medic, an aid bag, a backboard, and a dedicated vehicle and driver are on site during all training.
- Ensuring positive communication with aircraft is maintained throughout training.

5-13. PILOT IN COMMAND

The aviation unit providing helicopter support for the FRIES training and operation appoints the PIC. The PIC is responsible for all aspects of the flight and ensures—

- Aircrew members are current and qualified to conduct FRIES operations
- Aircrew members are informed of and understand their responsibilities in fast-rope operations.
- Procedures for planning, preparation and execution are adhered to IAW this TC, USASOC Regulation 350-6, and applicable SOPs and policies.
- Aircraft is properly configured to perform the mission.
- Aircraft is at the proper altitude, airspeed, and location as briefed.
- The command to deploy ropes is not given until the aircraft is at a stabilized hover.
- The aircraft position is maintained to keep ropes in contact with the surface until all descending ropers are on the ground (or extracting ropers are securely attached for extraction).
- Ropes are fully recovered inside the aircraft (or jettisoned) before the aircraft departs the stabilized hover position at the infiltration site.
- All personnel are briefed on in-flight emergency and safety procedures.
- Ropes are never deployed with anything other than night illumination attached to the free end.
- All personnel are briefed on fire support to be provided by the aircraft including:
  - — The nature of fire support the aircraft can provide.
  - — The time the fire support starts, shifts, and stops.
  - — The primary and alternate commands and signals that start, stop, lift, or shift fires.

5-14. FRIES MASTER

Units conducting FRIES operations will designate one overall FRM to organize, coordinate, and supervise the activities of the day, and AFRMs as needed. AFRMs are additional FRM-qualified soldiers who assist the primary FRM. An AFRM will be designated for each roping point. AFRMs are responsible for all preparation, inspection, and command and control of all roping activities on their points. FRMs also must—

- Ensure all FRIES roping personnel, including FRMs, AFRMs, and safeties, are qualified and current and understand the proper roping procedures for FRIES operations.
- Perform safety and serviceability checks on all FRIES equipment and rigging.
- Ensure the attachment bar or points are serviceable and free of any defects or contamination and that the quick-release mechanisms and safety pins are present and serviceable and operate correctly.
- Inspect the rope to make sure it has no contamination, damage, or defects that could make it unsafe.
- Check the rope to make sure it is the correct type and length for the operation (smooth rope for infiltration or looped for exfiltration).
- Ensure the rope is properly attached with safety pin in place and back coiled.
- Ensure the extraction harnesses are serviceable and properly worn.
- Ensure the rope chemlights are correctly rigged and illuminated when needed (two at the mount, two at the end, and two 15 feet from the end).
- Confirm the proper seating arrangement, including seat belts or a safety strap for ropers.
- Coordinate with the PIC and brief the aircrew, safeties, and ropers on the following:
  - — The aircraft flight, approach, flare, and position over the target site.
  - — The height of the aircraft above the target.
  - — A description of the FRIES site (rooftop, small clearing) and size, as well as the actions of the ropers upon arrival at the site.
  - — The signal to deploy the rope, who will deploy it, and how it will be done.
  - — Time warnings and actions at each time warning.
  - — Arm-and-hand signals.
  - — The use of radios and intercoms.
- Brief the PIC and crew chief on verbal commands and final adjustments over the target.
- Confirm the correct target and inform the pilot before the deployment of ropes or ropers.

WARNING

The FRM must ensure that nothing, other than rope illumination, is ever attached to the free end of the FRIES rope during deployment and that equipment is never attached and dropped.

- Deploy the ropes when the PIC gives the command DEPLOY ROPES. (This action may be delegated to the aircrew or safety.)
- Ensure safeties and AFRMS understand the time and activity sequence and all arm-and-hand signals.
- Ensure safeties know the proper procedures to—
- Keep the aircraft on target.
  - — Clear the ropes.
  - — Recover the ropes.
  - — Jettison the ropes.
- Observe the target and pass to the PIC, through the crew chief, final aircraft position adjustments over the target to make sure conditions are safe.
- Deploy the ropes when the PIC gives the command DEPLOY ROPES. (This action may be delegated to the safety.)
- Ensure all ropers know how to hook up for extraction and know the correct arm-and-hand signals.
- Direct the aircrew or safety to inform the pilot when the aircraft is on target (the PIC replies with the command DEPLOY ROPES).

CAUTION

When equipment is to be lowered, it should be lowered before the FRIES is deployed or personnel begin roping. When simultaneous, equipment and ropers will exit from opposite doors of the aircraft.

- — The safety repeats, " DEPLOY ROPES." The FRM signals the AFRMs to deploy ropes and makes a final check, ensuring all ropes from his aircraft reach the surface and are safe for descent (at least 15 feet of rope must be on the surface).
- — When conditions are safe, the FRM commands "GO" to the ropers and points to the rope or leads the stick out.

Note: If the FRM leads the stick out, the AFRM automatically assumes control. He completes marshaling the remaining ropers and is the last man to exit.

- Coordinate actions with the crew chief for the positioning or storage of seat belts, safety straps, rucksacks, door bundles, and communication cords.

Note: During night operations, upon completing descent, the AFRM observes the the ropes of his aircraft. When he sees that all ropes are clear, he issues the prearranged ALL CLEAR hand signal to the aircrew safety. This hand signal tells the aircrew that they are free to jettison or recover the ropes, and fly away.

5-15. FRIES AIRCREW MEMBER AND SAFETY

FRIES safeties are aircrew members who are FRM-qualified and perform the additional duties of the safety during helicopter roping operations. Safeties assist FRMs and pilots as needed. Aircrewmen performing safety duties differ from FRMs and AFRMs in that safeties are members of the aircrew (crew chief or gunner) whose orientation is the flying portion of the operation. Conversely, AFRMs are FRM-qualified members of the ground unit who are performing FRM duties as assistants to the primary FRM and who exit with the ground unit. Safeties remain in the aircraft and operate the doors, extend and retract the rope mount bars, and wear night vision goggles (NVGs) when appropriate. Safeties must have completed FRM training IAW this TC and the aircrew training program (ATP).

a. During roping operations, the safety's four major areas of responsibility are:

(1) Observe all activities in the aircraft and assist the pilot and FRM or AFRM as needed.

(2) Serve as the communication link between the PIC and the FRM. The safety monitors aircraft communication and relays information between the PIC and the FRM, constantly keeping both informed.

(3) Assist the FRMs and AFRMs in lowering cargo and clearing the aircraft of ropers.

(4) Clear and jettison the ropes, when appropriate.

b. The safety performs the following duties:

(1) Relays time warnings, aircraft positioning commands, and other information to assist the FRM.

(2) Conducts aircraft equipment rigging and serviceability inspections, making sure mounts, doors, and ropes are ready and safe for operations both pre-flight and en route.

(3) Verifies that the aircraft is at a stabilized hover with ropes in contact with the surface and remains so (or stops stick until corrections are made).

(4) Checks the rope to make sure 15 feet of rope are on the target surface.

(5) Tells the PIC, "ROPES DEPLOYED", when appropriate.

(6) Tells the PIC, "ROPERS OUT", when the first roper has exited.

(7) Helps the AFRM in controlling and spacing each roper's exit.

(8) Monitors the rope and ropers to make sure they do not get hung up, dragged, or dropped off site.

(9) Observes the exit of the last roper until the roper reaches the surface, moves clear of the rope, and gives the ALL CLEAR signal.

(10) Tells the PIC, "ROPERS DOWN", (station name: aft, forward, left, right) when all ropers are on the surface and conditions are safe to jettison the ropes.

(11) Passes the signals as required to position the aircraft, including HOLD, MOVE LEFT, RIGHT, FORWARD, BACK, UP, or DOWN.

(12) Acts as belay man during cargo lowering when needed.

(13) Jettisons ropes upon receiving the command JETTISON ROPES from the PIC. Tells the PIC, "Ropes are jettisoned," when the ropes have been recovered or jettisoned.

(14) Tells the PIC when ALL CLEAR.

CAUTION
This command tells the pilot the aircraft is cleared for flight.

(15) Deploys the rope for extraction upon receiving the PIC command DEPLOY ROPES.

(16) Observes the extracting rope and personnel and informs or guides the PIC through aircraft positioning and throughout the recovery flight.

5-16. FRIES ROPER

Ropers are responsible for notifying the FRM, AFRM, safety, or pilot if they observe any unsafe act or condition. Ropers may halt or call for a halt of roping for safety at any time. During FRIES insertion training, the number of ropers on the fast rope at one time is limited to three. During extraction training, the total weight per extraction bar or rope will not exceed a total of 1,500 pounds. When rucksacks are worn by ropers, the rucksacks must not exceed a weight of 35 pounds (total roper equipment weight will not exceed 50 pounds). Ropers will—

- Keep hands at head level.
- Maintain visual contact with lower ropers during their descent.
- Maintain a minimum of a 1-second interval on exit to avoid collisions.
- Keep at least two points of contact on the rope (both hands) at all times.
- Use their feet for additional breaking any time needed.
- Execute descents at a safe speed.
- Slow the rate of descent halfway down the rope to avoid landing on each other.
- Move quickly away from the rope upon arrival on the surface.
- Know correct wear of extraction (STABO) harness and extraction procedures.

WARNING
The FRM, AFRM, and safety must make sure all personnel are on the surface and have moved clear of the ropes before aircraft fly away or ropes are jettisoned.

Section V. FRIES RIGGING OF AIRCRAFT

The aviation unit is responsible for preattachment inspection and installation of FRIES anchor point mount bars and associated equipment. All aircraft used to conduct FRIES operations with US Army personnel must be rigged and equipped IAW current parent service directives--for Army aircraft, USASOC, SOAR, and TAPO directives apply. Rigging of the aircraft includes installation and removal of the seats and seat belts (or personnel restraints), and attaching and detaching the ropes.

Figure 5-1. UH-60 helicopter

> **CAUTION**
> **Army aircraft that have the FRIES mounts installed must have an airworthiness release for FRIES operations from the Aviation Troop Support Command (ATCOM) TAPO.**

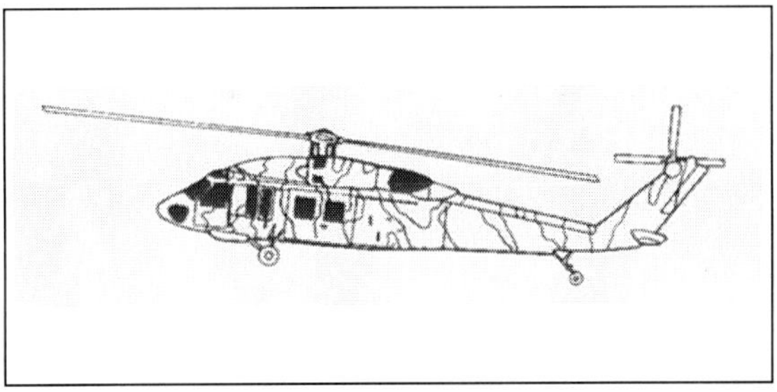

5-17. RIGGING OF FRIES IN UH/MH-60

The UH-60 is a twin-turbine, medium-speed, single-main-rotor helicopter (Figure 5-1).

The aviation aircrew or safety is responsible for—

- Inspecting the FRIES mount bar assembly, attachment hardware, and associated equipment to make sure they are serviceable and ready to use.
- Ensuring the bar assembly is correctly mounted to the aircraft (Figure 5-2).

Figure 5-2. Bar assembly correctly mounted to aircraft.

- Ensuring the mount is complete, functional, and free of any defect.
- Ensuring the mount and the aircraft floor are clean and free of any petroleum, oils, or lubricants (POL) or other contaminants that could weaken the rope or cause personnel to slip or fall.
- Ensuring the mount and aircraft are free of sharp edges or burrs that could cut the rope.
- Ensuring no metal fatigue or structural weakness exists that could cause the system to fail.
- Extending locking and retracting mount bars as needed.
- Lowering and belaying equipment as needed.
- Opening, locking, and closing cargo doors as needed.

Note: For cold-weather operations or during flights of long duration, the cargo doors are closed and latched until the time specified for opening. The aircrew will operate doors or directly supervises those who do. In good weather, doors should remain locked open.

- Removing or reconfiguring troop seats.
- Providing seat belts or floor-mounted personnel restraints for ropers' use during flight.
- Attaching and detaching ropes as needed.

Note: Clusters of rucksacks or small door bundles (weight limited to 250 pounds) may be lowered (belayed) on rappelling ropes. Cargo lowered this way must never be free dropped (the belay must start at the time the bundle is pushed [tipped] out the door).

5-18. RIGGING OF FRIES IN CH/MH-47

The CH-47 is a tandem-rotor, medium-transport helicopter (Figure 5-3).

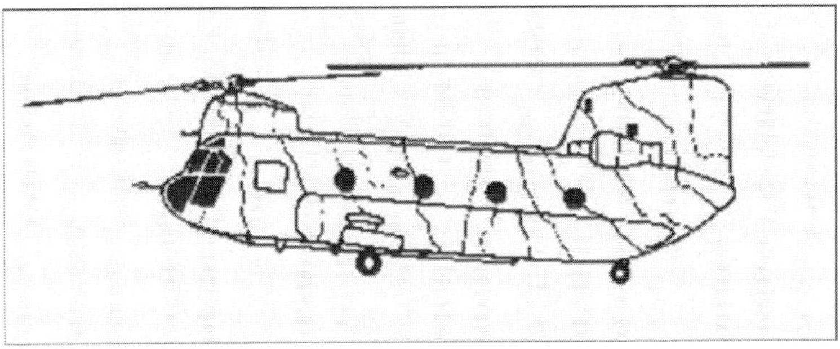

Figure 5-3. CH-47 helicopter.

The aviation aircrew or safety is responsible for:

- Inspecting the FRIES mount bar assembly, attachment hardware, and associated equipment to make sure they are serviceable and ready to use.
- Ensuring the bar mounts are correctly mounted to the aircraft (Figure 5-4, rear; Figure 5-5, winch).

Figure 5-4. Bar mounts correctly mounted, rear.

Figure 5-5. Bar mounts correctly mounted, winch.

- Ensuring the mount is complete, functional, and free of any defect.

- Ensuring the mount and the aircraft floor are clean and free of any POL or other contaminants that could weaken the rope or cause personnel to slip and fall.
- Ensuring the mount and aircraft are free of sharp edges or burrs that could cut the rope.
- Ensuring no metal fatigue or structural weakness exists that could cause the system to fail.
- Extending locking and retracting mount bars as needed.
- Lowering and belaying equipment as needed.
- Opening, locking, and closing cabin doors as needed.
- Removing or reconfiguring troop seats.
- Providing personnel restraints for ropers' use during flight.
- Taping and padding any sharp edges that can come in contact with ropes or personnel
- Attaching and detaching ropes as needed.

5-19. RIGGING OF FRIES IN MH-53

The MH-53 is a USASOC twin-engine, single-rotor, medium-transport helicopter used by the Army (Figure 5-6).

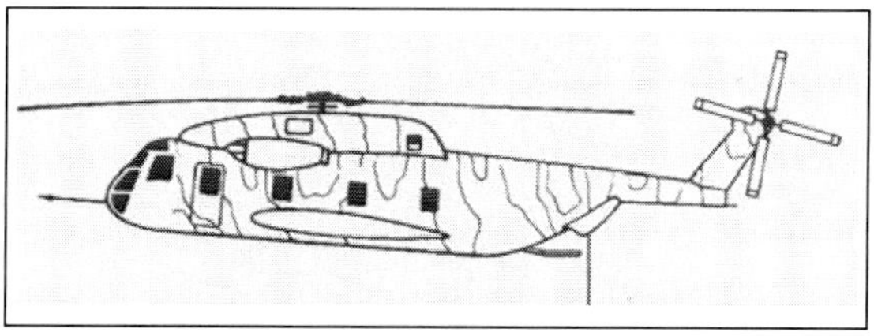

Figure 5-6. MH-53 helicopter.

The MH-53 primary FRIES mount bar assembly is located overhead, just aft of the rear ramp. The ramp mount bar assembly spans the cabin width and has a two-pin release mechanism on each end. The ramp position mount assembly provides two points; each will accommodate one FRIES rope or one cargo lowering rope. The ropes are carried back-coiled on the ramp.

a. The alternate position for fast-rope descent from the MH-53 is the crew entrance in the right side of the fuselage immediately behind the pilot's position. In this position, the fast-rope mount must be connected to the external hoist point. The FRIES master is responsible for inspecting the MH-53 mount system for its presence, security, and operability.

b. The aviation unit is responsible for—

- Installation and maintenance of the fast-rope mount assemblies at each position to be used.

- Configuring the aircraft to accommodate the number of soldiers and type of mission planned.
- Ensuring seat belts or restraint straps are present to secure ropers during flight.

Note: The MH-53 aircrew can improvise handrails to assist roper exits. When used, improvised handrails will be rigged IAW USAF special operations SOPs.

- Configuring the MH-53 for fast-roping when equipped with external 650-gallon fuel tanks.
  - Remove the ramp door to install the cross bar.
  - Place the bar in the brackets with the two quick-release paddles facing forward.
- Configuring aircraft armament (50-caliber machine gun and the GAU-2B ramp minigun) for FRIES operations.
  - When using the aft right roping position, point the 50-caliber machine gun to the 6 o'clock position and secure the gun with bungee cords.
  - When using the left aft rope, disconnect the minigun ammunition chute and stow it in the ammunition can, or disconnect it at the ammunition can and securely wrap it around the minigun. Use bungee cords to secure the ramp minigun in the firing position facing the 6 o'clock position.

Note: Ideally, only the right aft rope is used when the GAU-2B ramp minigun is installed. Although not recommended, the left aft rope may be used, if required.

Section VI. EQUIPMENT MAINTENANCE AND INSPECTION

Before conducting a FRIES operation, the FRM must inspect all FRIES equipment (ropes, STABO harnesses, and mount bars) for serviceability and readiness for use.

> **WARNING**
>
> **FRIES ropes and harnesses are critical life support equipment and must be stored, cared for, and maintained as life support items.**

5-20. ROPES AND HARNESSES

FRIES equipment should be maintained in the same manner as a parachute. The unit rigger section can provide detailed guidance on appropriate inspection, care, and maintenance of FRIES equipment.

a. FRIES ropes and harnesses must always be stored in a clean, cool, dry space out of direct sunlight and free of chemicals or chemical vapors. Equipment that becomes wet with fresh water should be hung up to dry (indoors) on hardwood pegs. Equipment that is exposed to salt water or becomes imbedded with dirt or mud should be washed and rinsed in fresh water (within 72 hours) and hung up to dry (indoors) on hardwood pegs (out of direct sunlight).

b. Before conducting a fast-rope operation, the FRM—

- Inspects the fast rope thoroughly and carefully. Checks the rope length to ensure it is the correct rope for the operation planned.
- Checks the woven loop on the mount end for excessive wear or chemical contamination. Checks the rope along its entire length for fraying, cuts, and chemical contamination.
  - Do not use a rope that is severely frayed. (Light fraying on the rope from normal use does not weaken the rope.)
  - Do not use a rope when any single strand is cut halfway through or has two or more cuts that penetrate one-third or more through any strand's thickness within 1 foot of the running length of the FRIES.
- Inspects the rope for contamination of acid, alkaline compounds, saltwater, fire extinguisher solutions, or petroleum-based solvents. Changes in color caused by chemicals are usually blotchy and have an unusual odor. Although used ropes gradually change color, such changes do not indicate a decrease in strength unless the change is due to contact with strong chemicals. Changes occurring because of use are usually uniform throughout the length of the rope.

WARNING

Any rope known or suspected to be chemically contaminated, cut beyond allowable limits, excessively frayed, or sun damaged must be taken out of service, marked for destruction, and turned in to the rigger section.

- Inspects the extraction loops to the same standard as the main rope. Ensures the woven attachment loops are secure.
- Inspects the harness to ensure—
  - Ropers are wearing the harness under all load-carrying equipment.
  - Ropers have properly fastened all connectors.
  - Harness material and stitching are not cut, torn, or contaminated, and all hardware is free of corrosion and is in operable condition.

Note: Serviceability inspection and maintenance of the extraction harness are the responsibility of the rigger-qualified personnel.

5-21. FRIES MOUNT BARS

The aviation unit is responsible for the installation, removal, storage, and maintenance of FRIES mount bars. The US Army Special Operations Aviation Regiment, Ft Campbell, KY, prepares and provides checklists for these procedures. Aviation units participating in FRIES operations must have and comply with current SOAR guidance.

Aviation units:

- Install and maintain the FRIES mount bar; check mount bolt torque and witness marks.
- Check the FRIES mount bar for security and operability.
- Check the condition of the mount bar; inspect for bends, dents, distortion, cracks, corrosion, and contaminants.
- Ensure the fast-rope quick-release pins are installed and operable (two each side).
- Check the security and condition of the rope-release system, as follows:
  - The mounting bolts are installed and properly torqued.
  - The safety pins are installed, and lanyards are present.
  - The system locks and unlocks as designed.
- Ensure the release cables are attached at both ends and are not broken or kinked.
- Check the release handles to ensure they are securely mounted and can move freely.
- Check the latch as follows:
  - Latch or close the release mechanism and apply a downward pressure on the release arm. Attempt to pull the release handles the release should not open.
  - Relax pressure on the release mechanism and pull the handles the release should open.
  - With no pressure on the release mechanism and with the safety pin installed, attempt to pull the release handles--the release should not open.

Section VII. OPERATIONAL REQUIREMENTS AND LIMITATIONS

This section establishes medical coverage and communication support requirements for FRIES operations, including during adverse weather conditions and night operations.

5-22. MEDICAL SUPPORT

Dedicated medical support is required for FRIES training and demonstrations. A qualified medic with assistant, aid bags, two backboards, two stretchers, and a dedicated vehicle with a driver must be present to respond to any mishaps. The nearest medical evacuation (MEDEVAC) unit will be notified of where and when FRIES training is scheduled. MEDEVAC response will be coordinated as on call. Training or demonstrations will not be conducted without medical support present. The ground force commander may direct the FRIES aircraft to transport life or death (urgent) patients when no other MEDEVAC unit is available, providing the medical team is trained and properly equipped with the appropriate equipment for aerial transport of patients. This must be included in the air mission briefing prior to training and coordinated with the medical team providing coverage. FRIES operations that are evaluated as high risk or that occur at night must have a certified combat lifesaver on location and medics (equipped as stated above) in the immediate vicinity.

5-23. COMMUNICATIONS REQUIREMENTS

During peacetime training and operations, FRIES lift aircraft and ground elements will maintain continuous radio communications. Aircrew personnel clearing ropes at each FRIES station must be on the intercom system, monitoring aircraft internal communications. They must also be proficient in the use of arm-and-hand signals as precoordinated between FRIES users and aircraft crews. The designated aircrew member will relay time warnings between the PIC and FRM. Routine time warnings are 10-, 6- and 1-minute warnings to the FRM. Ground-to-air communications must be continuously maintained throughout FRIES training.

5-24. ADVERSE WEATHER OR TERRAIN CONDITIONS

FRIES training and operations will ***not*** be conducted under the following conditions:

- Anytime temperatures are below 32 degrees Fahrenheit or wind-chill factors that could cause cold-weather injuries exist including those caused by helicopter rotor wash or extraction cruise speed.
- Anytime water or ice on the rope prevents personnel from controlling their descent.
- Anytime the rope is exposed to freezing conditions for a sufficient length of time to freeze, thereby reducing its tensile strength.
- Anytime dust out or white out conditions restrict the aviators or the ropers vision.
- Anytime trees or vegetation on the LZ is too dense for safe training.
- Anytime the LZ is too steep or too slippery for safe training.
- Anytime winds exceed 30 knots or are too gusty for safe operations.

5-25. NIGHT OPERATION REQUIREMENTS

Night operations require the following procedures.

a. Attach two chemlights at the free end of the rope and two above the extraction attachment loops (15 feet from the end of the rope) to aid in determining when adequate rope is on the surface.

b. Attach two chemlights (one primary and one backup) to the loop of the rope, or to the rope mount, to guide personnel exiting the aircraft to prevent glare from obscuring vision.

c. Attach one chemlight or luminous tape to the top of each roper's helmet. If a chemlight is used, select a color different from the colors used on the ropes.

CAUTION

Ropers will not wear NVGs during descent. NVGs limit depth

perception and cause a tunnel-vision effect. Static FRM and Safeties may wear NVGs to help confirm that fast ropes are fully deployed. The aircrew will wear NVGs as required by flight mission needs.

Section VIII. FRIES PROCEDURES

The risks of personal injury and equipment damage during FRIES operations can be reduced by detailed planning, thorough training, and strict adherence to safe procedures. Conducting FRIES operations with heavy loads requires personnel to be proficient in these operations. FRIES users must limit ropers to no more than 6 descents with combat load and no more than 10 total descents in any 24-hour period. The total weight of equipment soldiers may carry during training is limited to never more than 50 pounds. This section provides standardized procedures for FRMs, PICs, and ropers. These procedures are common to all units and aircraft, except as noted.

5-26. FRIES MASTER

The FRM performs the following series of actions.

- Briefs members of his team and aircrew.
- Inspects team members and equipment.
- Installs the FRIES rope in the aircraft and conducts safety checks.
- Relays time warnings to team members. (Time warnings are a tool to help keep aircrew and ropers' actions synchronized and can be modified according to user needs; however the 1-minute warning should always be used.)
- Breaks chemlights, if required. Two chemlights or luminous tape will be attached at the anchor point to identify the handhold point.

WARNING

Equipment must never be attached to the FRIES and dropped. As little as 10 pounds of weight can exceed the load limit of the FRIES mounts when the load stops at the end of a 60-foot drop. The only items to be attached to the fast rope during deployment are the position markers.

- Ensures the rope is properly configured for deployment (back coiled to prevent tangles).
- Ensures the team members are in order of exit no later than the 1-minute warning.
- Confirms target on final approach.
- Deploys the rope and ensures it reaches the ground. During night

operations, the rope is marked with six chemlights: two at the mount, two on the end, and two 15 feet from the end.

- Deploys personnel.
- Accounts for personnel, and signals aircrew.

> **WARNING**
> **Because the FRIES mount in the CH-47 helicopter is narrow, special care must be taken by ropers to avoid fouling weapons or equipment on the frame during exit.**

5-27. PILOT IN COMMAND

The PIC briefs the FRM, safety, and roping personnel before FRIES operations.

a. The PIC briefing includes:

- Approaching, loading, unloading, and departing the aircraft.
- Actions in the aircraft.
- Flight route and checkpoints en route.
- Altitude or height for roping.
- Time warnings to FRM as needed (recommend 10 minutes, 6 minutes, and 1 minute).

b. When the PIC is over the intended objective he calls, "DEPLOY ROPES" Fast-rope warnings and expected aircrew coordination calls are as follows:

(1) The aircrew or FRM provides any final position adjustments of the aircraft with the commands of MOVE (left, right, forward, back, up, down, or hold) as needed to place the ropes ON TARGET. (When the pilot is to hold the position, an explanation will be provided as soon as possible.)

(2) When the FRM deploys the ropes, the aircrew member calls, "ROPES DEPLOYED."

(3) When the first roper exits the aircraft, the aircrew member calls, "ROPERS OUT."

(4) The safety watches the last roper until he is on the ground and gives the prearranged signal. The safety then tells the PIC, "ALL ROPERS AWAY," and identifies his station. This alerts the PIC that the safety is going to either jettison or recover the ropes.

(5) The safety will then either recover or jettison the rope as planned, clear the aircraft all around for obstructions, and call, "ROPES CLEAR (station)" This informs the PIC that the aircraft is clear for flight.

c. The PIC also briefs on emergencies that could occur in the following situations:

- En route (when ropers are inside the aircraft).
- Before deployment of ropes.

- When fast-rope personnel are on the ropes.
- After the ropers are on the surface.

5-28. ROPERS

Ropers perform the following actions:

a. At the command STAND BY (given at 1-minute warning), ropers make final check of themselves and take the prepared-to-exit position.

b. At the command GO, the first roper exits the aircraft, rotates his body 90 to 180 degrees to make sure his equipment clears the aircraft, places the fast rope between the arches of his feet, and begins descent. The roper uses his hands and feet to slow his descent approximately two-thirds of the way down.

c. Subsequent ropers exit at 1-second intervals using the same procedure, but begin slow descent approximately half-way down to avoid landing on other ropers.

d. During descent, ropers keep a lookout and break as necessary to avoid landing on obstructions or on fellow ropers.

e. Ropers prepare to land just before reaching the ground by spreading their legs about shoulder-width apart with their knees slightly bent.

f. At landing, ropers quickly move clear of the ropes to avoid collisions with descending ropers.

CAUTION

According to AR 95-3, anytime operations occur over water, all personnel will wear an approved personal flotation device as identified in FM 57-220 or USASOC Regulation 350-2.

5-29. EQUIPMENT-LOWERING PROCEDURES

Anytime a roper's load exceeds a total of 50 pounds or his rucksack weight exceeds 35 pounds, equipment will be lowered (as clusters or bundles) on separate ropes. Lowering equipment from heights great enough to cause equipment damage necessitates belaying the equipment during descent to avert damage. Belaying devices used in mountaineering allow ropers or safeties to control equipment descent.

Figure 5-7. Rucksacks upright on cargo deck.

CAUTION

When equipment is to be lowered, it should be lowered before the FRIES is deployed or personnel begin roping. When done

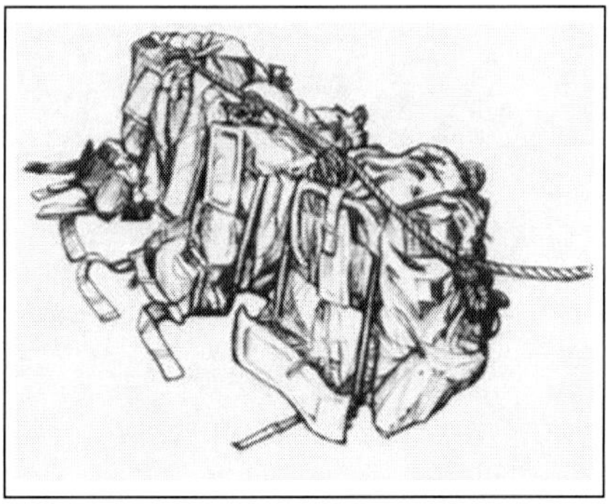

simultaneously, equipment and ropers will exit from opposite doors of the aircraft.

a. Rigging equipment to be lowered requires rigging rucksacks in clusters. Clusters of up to five rucksacks may be lowered (smaller clusters work better). FRMs will ensure the total weight of any cluster does not exceed 250 pounds.

(1) Place the rucksacks upright on the cargo deck of the helicopter (Figure 5-7) .

(2) Lay a sling rope across the frames between the shoulder strap attachment points.

(3) Tie a bowline knot at each end of the sling and a figure eight knot at each rucksack.

(4) Attach a snaplink to the frame of each rucksack, to the loops in the sling rope, and to the end of the lowering rope.

(5) For Lowe rucks, install the issue H-harness on the rucks and join them into clusters with a sling rope tied to the quick-release attachment buckles.

b. Rigging of the lowering rope to belay the rucksacks requires the following actions.

(1) Backcoil a rappel rope on the helicopter floor at the gunner=s position.

(2) Attach a snaplink and belay device (either a figure eight [Figure 5-8] or an SMC ladder [Figure 5-9]) to the floor tiedown (with a snaplink) inboard of the ceiling rappelling anchor point.

(3) Attach a snaplink to the ceiling rappelling ring.

(4) Route the belay rope from the backcoil, through the belay device, then up through the ceiling snaplink, and down to the cluster snaplink.

(5) Tie the end of the belay rope into a bowline with halfhitch and snap it into the cluster snaplink.

(6) The FRM identifies two ropers to push the cluster or bundle out of the aircraft. He also identifies an AFRM or a safety to belay the cluster or bundle. The FRM briefs these personnel on the time to lower the bundle, the signals to be used, and the procedure to clear the belay system before deploying the FRIES. In performing his

belaying duties, the AFRM or safety must The aviation aircrew or safety is responsible for:

- Provide sufficient slack to allow the pushers to eject the cluster from the aircraft.
- Control the descent until the cluster reaches the surface.

Figure 5-8. Belay Device figure eight SMC

Figure 5-9. Belay device ladder

- When the cluster is on the surface, unsnap both the upper and lower snaplinks, drop the remainder of the rope from the aircraft, and immediately look out to make sure conditions are safe to deploy the FRIES.
- Inform the FRM and pilot when conditions are safe.

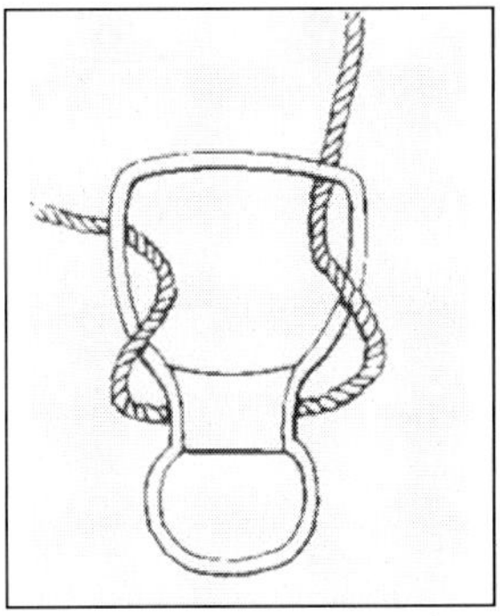

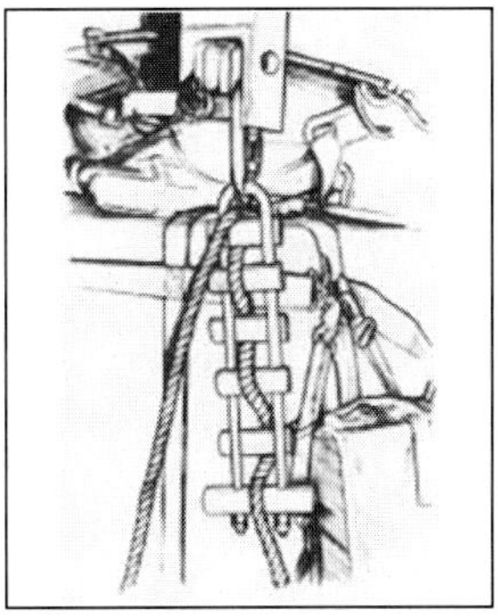

- Deploy FRIES on command

Note: Expect to reposition aircraft before ropers exit to avoid ropers landing on equipment.

5-30. HELICOPTER OPERATIONS FRM CHECKLIST

Table 5-1 shows a checklist of duties of key personnel in fast-rope operations.

PRE-ROPING ACTIONS
Receive a briefing from the OIC or the AMC. Coordinate and brief all participants. Rig aircraft and conduct a joint inspection. Brief ropers, AFRMs, and safeties. Rig and inspect ropers.

Conduct a static rehearsal.
AIRCRAFT LOADING Position equipment and personnel. Make sure all personnel have straps or seat belts. **ACTIONS IN FLIGHT** Monitor the command net. Monitor the aircrew net. Monitor the flight route. **ACTIONS AT 10-MINUTE WARNING** (applies to long infiltrations) Issue the 10-minute time warning and GET READY. Check equipment and belay system hookup. Check fast-rope hookup. Secure fast-rope bar in position. Make sure rope is back coiled and markers are attached. **ACTIONS AT 6-MINUTE WARNING** Issue the 6-minute time warning. Remove personnel restraints or seat belts. Position personnel and equipment. Break chemlights (night operations). Open aircraft doors, if required. **ACTIONS AT 1-MINUTE WARNING** Issue the 1-minute time warning and STAND BY. Position ropers in stick formation. **ACTIONS AT FLARE** Identify the target area. Deploy bundles/equipment (Safeties) and clear ropes. Deploy FRIES ropes (FRM and AFRM). Check ropes to make sure 15 feet of rope are on the surface (FRM).

Table 5-1. FRM helicopter fast-rope operations checklist

ACTIONS FOR DESCENT
~ FRM, AFRM, or safety positions the number 1 man at the rope. (The FRM may exit first or last.) ~ FRM issues the command GO. AFRM echoes GO command, and the ropers exit the aircraft. Safety informs PIC, "ROPERS OUT."

AFRM or safety controls the rate of exit of the ropers. AFRM exits last. Aircrew or safeties observe the last roper. Safety tells the PIC, ALL ROPERS AWAY, after the last roper is on the surface and signals. PIC issues command JETTISON or RECOVER ROPES. Aircrew or safety jettisons or recovers ropes, and issues ROPES CLEAR report to the pilot.

Table 5-1. FRM helicopter fast-rope operations checklist (continued).

5-31. FRIES COMMANDS AND SIGNALS

The following standardized arm-and-hand signals are provided for FRIES operations. Ground-to-aircraft arm-and-hand signals are military standard as found in the Pathfinder field manual 57-38.

a. Time warnings (10-, 6-, and 1-minute) are given by holding up the appropriate number of fingers and verbally sounding off.

b. The signal for GET READY is the same as the standard Airborne signal—arms extended horizontally, fingers extended vertically and joined.

c. The signal for STAND-BY is arms extended downward at a 45-degree angle from the body with hands closed except for index fingers, which are extended, pointing to the floor

d. To execute the signal for DEPLOY ROPES, the FRM or safety forms both hands into fists at mid-thigh level and sweeps arms horizontally outward. At full extension, he opens his hands as if dropping the ropes out of the helicopter and sounds off with DEPLOY ROPES.

e. The signal for GO is arms raised to horizontal, hands closed with index fingers extended pointing to the ropes.

f. The signal for STOP or ABORT is hands closed into fists with arms raised across the forehead .

g. The signal for SIT-DOWN and DON SEAT BELTS is pointing to the floor, followed by moving both fists around waist, as if donning a belt.

h. The signal for OK is hands raised to eye level, index finger and thumb forming a circle with other fingers extended.

Section IX. EMERGENCY ACTIONS

Procedures for correcting an emergency condition that could reasonably be encountered are described in this section. Multiple emergencies, adverse weather, or other unusual conditions may require modification of these procedures. The nature and severity of the emergency dictate the response necessary; therefore, personnel must use sound judgment in determining the correct action to be taken.

5-32. EMERGENCIES BEFORE ROPING STARTS

All personnel sit down, don their seat belts, and take further instructions from the pilot or crew chief.

5-33. EMERGENCIES AFTER ROPING STARTS

Emergency procedures to be followed after roping starts are discussed in this paragraph.

a. **Unsafe Drift or Premature Liftoff.** Anyone who detects the aircraft having drifted off the site must immediately stop training and inform the FRM, AFRM , safety, or PIC.

(1) The FRM, AFRM, safety, or roper stops stick.

(2) The ropers stop descent and lock in.

(3) The FRM or crewman informs the PIC, and guides him in moving the aircraft back on target.

(4) The unit continues operations.

b. **Rope Hung or Snagged (On Ground, Tree, Building).** Anyone who detects the rope being snagged or hung must immediately stop training and inform the FRM, AFRM, safety, crew chief, or PIC.

(1) The safety ensures ropers are off the rope and are clear.

(2) The aircraft descends or lands as needed.

(3) The safety frees the rope and informs the pilot.

(4) The unit resumes the operation.

c. **Lost Communications.** All training and operations must be planned to include the use of the intercom between the PIC or crewmen and the FRM. If the intercom fails, the following arm-and-hand signals can be used until the rope can be cleared and the intercom is restored.

- Wrists crossed diagonally above the forehead with clenched fists to signal stop stick.
- Finger pointed toward the exit and a verbal GO to signal ropers to exit.
- Open palm moved and faced in the required direction to signal aircraft movement.
- Clenched fist to stop aircraft movement.

Note: Training will not be conducted without radio communications between the pilots or aircrew and the FRM and the ground.

Section X. SAFETY

Safety is everyone's responsibility. All personnel involved in FRIES operations are responsible for identifying hazardous situations and preventing injuries to personnel. Anyone who observes an unsafe condition or act is authorized to halt the operation and inform the FRM or the PIC.

5-34. PREFLIGHT AND INFLIGHT

Anyone who detects an inflight emergency will immediately notify the PIC, crew chief, or safety and follow aircrew instructions.

a. Before a FRIES operation, the FRM will coordinate with the PIC both routine and emergency procedures as well as any unique conditions or potential problems. He also discusses the procedures to be used during flight, during roping, and before the aircraft's departure from the roping site. The FRM then discusses the information with all the AFRMs, safeties, and crew chiefs as well. He reviews the time warnings (including arm-and-hand signals) and appropriate actions such as who will deploy ropes, verify rope status, and perform rope-clearing procedures.

b. The mission prebrief covers any aircraft equipment that, if disturbed, could cause an aircraft or fast-rope problem. The problem areas should be made known to all personnel concerned and preventive measures taken to neutralize or eliminate them where possible.

c. The FRM gives the crew chief the final position adjustment over the target, using hand signals.

d. All safety straps and communications cords that are not in use are securely stowed. All safety straps and communication cords in use should be positioned so they do not present a hazard to personnel movement in the aircraft.

e. After the PIC gives the command DEPLOY ROPES, the FRM is responsible for deploying the rope and making sure it is in contact with the surface before ropers exit.

Note: The safety is responsible for observing rope and ropers and for making sure the pilot maintains aircraft position until all ropers are on the surface.

CAUTION

According to AR 95-3, anytime operations occur over water, all personnel will wear an approved personal flotation device as identified in FM 57-220 or USASOC Regulation 350-2.

5-35. DURING ROPING

The following safety procedures apply during roping.

a. During the actual descent, fast-ropers will not use NVGs because they limit depth perception and create a tunnel vision effect.

b. During descent, ropers must watch for obstructions and lower ropers, and must break as necessary.

c. Individual ropers will lock in during emergencies by wrapping one leg around the rope once or twice and standing on the fast rope with the other foot.

d. Ropers must maintain sufficient intervals on the rope to prevent contact with lower ropers. Individual ropers exit when the head of the man in front of them disappears below the ramp.

e. Ropers execute descents at speeds commensurate with their experience and proficiency in FRIES infiltration operations. (For example, less-experienced ropers should descend slowly until they become more proficient.)

f. The number of ropers on each rope at one time must not exceed the safe limit of the rope mount point in use, as follows:

(1) Do not exceed 1,500 pounds per rope on the MH/UH-60.
(2) Do not exceed 1,500 pounds per rope on the MH/UH-47.
(3) Do not exceed 1,500 pounds per rope on the MH/UH-53.
(4) Do not exceed 1,300 pounds per rope on the MH-60G.

5-36. SAFETY BRIEFING

All units must present a safety briefing before conducting FRIES training or operations. The safety briefing must cover the following (as a minimum):

- The planned event (day or night, tower or helicopter, high or low hover, insertion or extraction).
- Actions on the aircraft (or tower).
- Operation (or training) location and any hazards within 500 meters.
- Identification of key personnel, their location, and their duties.
- FRIES equipment and its characteristics.
- FRIES equipment rigging or inspection.
- Roping personnel preparation (shirttails tucked in, chin strap fastened, loose straps taped, gloves and goggles on).
- Arm-and-hand signals.
- Emergencies and emergency procedures.
- Medical procedures (medic with aid bags, backboards, litters, and dedicated vehicle with driver on site).
- Communications requirements (frequencies, call signs, required reports).
- Any other information essential to safety.

5-37. SAFETY REMINDERS

The following safety reminders will be reviewed by all key personnel before conducting operations.

a. Ropers must wear long-sleeve shirts or jackets, full-length pants, laced boots, helmets, protective goggles, and heavy leather gloves.

b. MILES harnesses must not be worn during fast-roping.

c. Ropers will not wear NVGs.

d. Commanders must personally approve the rope site selected for night operations. The FRIES site must be large enough to permit all ropers, upon reaching the surface, to move clear of the rope and for units to consolidate or re-form.

e. All night rope operations must be assessed as medium risk or higher.

f. FRIES descent training is not authorized with a fast rope longer than 90 feet.

g. Ropers must not carry more than 50 pounds of equipment during FRIES descents. The maximum weight of rucksacks is limited to 35 pounds. When rucksacks exceed 35 pounds (or total load exceeds 50 pounds), gear must be lowered (belayed) on a separate system, preferably from the opposite door.

h. Ropers will not carry bulky equipment that could interfere with FRIES operations. Equipment that cannot be carried by the soldier during roping (due to its weight or bulk) can be lowered (belayed) during insertion. During extractions, rucksacks must be attached to the extraction loops of the FRIES rope using snaplinks.

i. Ropers must not make more than 6 combat equipment descents or a total of 10 descents in a 24-hour period.

j. FRIES training must not be conducted if—

- The aircraft is unstable.
- The fast rope is not fully deployed (15 feet of the rope must be on the surface).
- Any obstacle is present that could interfere with the descent of the ropers.
- The rope is frozen or slippery
- Anyone identifies any unsafe condition.

k. Safety personnel must—

- Position themselves so they can observe active FRIES points with out obstructing ropers.
- Conduct final rigging checks of the mount, equipment, and ropes at the 10-minute warning.
- Ensure the aircraft is over the target area and the ropes reach the ground.
- Constantly keep both the PIC and FRM informed—
  - — When the aircraft is in position for deployment.
  - — When anything is being deployed from the helicopter (ropes, bundles or personnel).
  - — When 15 feet of rope are on the surface.
  - — When the first roper exits on the fast rope.
  - — When all ropers on the surface are clear.
  - — When the ropes are recovered or jettisoned.
  - — When the aircraft is cleared for flight.
  - — When 15 feet of the extraction rope is on the surface.
  - — When extraction personnel are attached to the rope and ready for lift off.
  - — When extracted personnel are safely lowered to the ground.
  - — When extracted personnel are out from under the aircraft.
  - — Any other conditions that could impact safe operations.

Note: If the FRM and the safety exit, the crew chief performs rope clearing and or jettisoning duties.

> **WARNING**
>
> **Using the word CLEAR tells pilots the aircraft is ready to fly. Before any use of the word CLEAR, FRMs and safeties must ensure that all personnel are well away from the ropes and the ropes have been fully recovered or released (during insertion), or all personnel are securely attached to the ropes (during extraction).**

l. FRIES training must not be conducted in densely wooded areas or areas prone to blowing dust, sand, or snow, which can produce a hazard for ropers or the aircraft.

m. FRIES training and operations must not be conducted during severe weather such as rain, sleet, ice, or extreme cold.

n. The possibility of personnel injury during FRIES training and operations is always present. Every soldier involved in FRIES activities must maintain constant vigilance to prevent personnel injury and equipment damage.

CHAPTER 6

EQUIPMENT

This chapter discusses the basic equipment used for rappelling operations (Appendix B). It covers the types of ropes, the procedures for rope selection, and the care for each rope type. It also discusses the two methods of coiling and the different snaplinks used to join the equipment and rope.

Section I. ROPES

Ropes are the most important items of equipment for the rappeller. They provide access down the obstacle or aerial platform, while ensuring individual safety.

6-1. TYPES OF ROPES

Two types of ropes are available to the rappeller—*static* and *dynamic*.

a. **Static Rope.** This rope allows for minimal stretch of the rope. The static rope stretches about 5 to 15 percent at the point of failure and about 2 percent under a working load. The minimum tensile strength for an 11-mm (7/16-inch) static rope for military use is 4,500 pounds. *Kernmantle* static ropes range from 3 millimeters to 11 millimeters in diameter and are constructed similar to the 550-parachute cord. (*Kern* means core and *mantle* means sheath.) The internal core of the rope is constructed of a continuous multifilament-nylon yarn, which is spun into continuous parallel strands. This internal core is then covered with a nylon-braided outer sheath. Due to the internal parallel strand construction of the static rope, there is less spinning and kinking than with a dynamic rope. Static ropes can be used for slings, harnesses, etriers (stirrups), rappelling, rope installations, hauling lines, and where stretch of the rope is not desired.

b. **Dynamic Rope.** This rope allows for stretch within the fibers of the rope, which can be a disadvantage in rappelling, prusik climbing, and other applications. Dynamic ropes are more susceptible to abrasion and wear. They have about a 5 to 10 percent working stretch. The minimum tensile strength for an 11-mm (7/16-inch) rope for military use is 4,500 pounds. Two types of dynamic ropes are nylon-laid and kernmantle:

(1) *Nylon-laid ropes.* Synthetic fibers are the best climbing and rappelling material. Nylon has become the standard material for climbing ropes and has replaced manila, flax, hemp, and sisal. The military mountaineering critical application rope is constructed in a *laid* fashion with continuous multifilament-nylon fiber yarn twisted into strands. Three strands are twisted into a climbing rope of a specific diameter (Figure 6-1). Nylon-laid ropes are easy to inspect for serviceability by twisting the fibers. Laid ropes tend to untwist slightly when under a load, causing kinking and spinning. They are also susceptible to abrasion. The following specifications pertain to the standard military nylon-laid climbing/rappelling rope:

- 36 1/2 meters (120 feet) long.
- 11 millimeters (7/16 inch) wide.
- At least a 4,500-pound tensile strength.
- Right-hand lay.
- One-third stretch factor (at point of failure).

- May lose as much as 15 percent of its strength when wet.
- Weighs 6 pounds when dry.

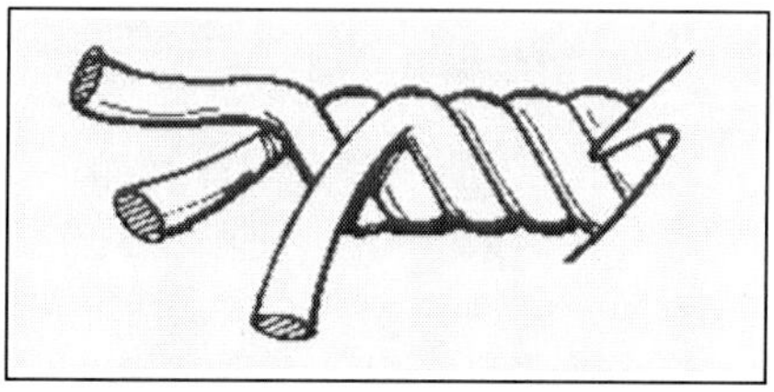

Figure 6-1. Nylon-laid rope.

(2) *Kernmantle ropes.* These ropes are similar to their static counterparts in that they consist of an inner core and an outer sheath. The core is constructed of continuous twisted nylon filaments, which are laid or braided together and enclosed in a tightly braided outer sheath. These ropes are well suited for climbing on rock, snow, or ice where a brief elastic stretch of the rope occurs during a fall. They have a stretch of about 40 percent at the point of failure. The breaking strength is high, and there are no exposed strands for rock crystals to work between to damage the rope (Figure 6-2). There is less sliding friction through the snaplink and over other surfaces, since the outer sheath is smooth. Kernmantle ropes are available in many sizes, lengths, stretch factors, tensile strengths, and fall ratings.

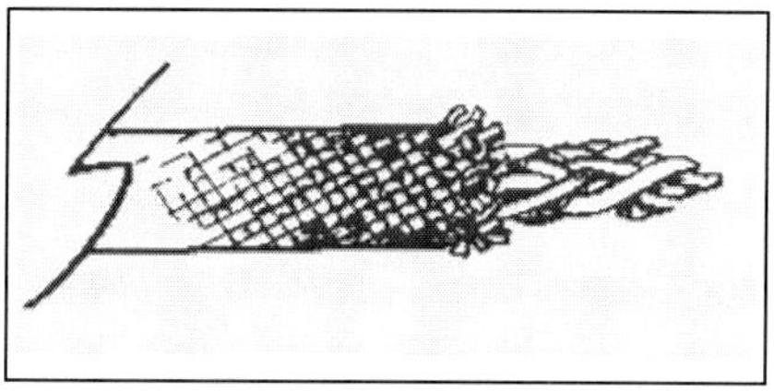

Figure 6-2. Kernmantle (core and sheath) rope.

6-2. ROPE SELECTION

Rope selection for military operations requires ropes of various types, lengths, and diameters; and depends on its intended use (static or dynamic), environment (water-repellant), climate, mission (lightweight ropes), and other factors. Static ropes should have minimal stretch; be smooth, flexible, and strong; resist abrasion and cutting; and have a low-impact force.

a. *Laid rope* is easier to inspect (by twisting the strands apart to observe the interior surfaces), has greater friction in holding knots, and is less expensive than kernmantle ropes. Its disadvantages include: much lower abrasion resistance, extremely high stretch factor for both static and dynamic loads, greater friction on rock, readiness to kink, tendency to spin a free-hanging rappeller, and greater absorption of water and dirt.

b. *Kernmantle* rope is easy to work with, has less friction, does not kink easily, and has less susceptibility to abrasion. The outer sheath protects the internal fibers from dirt, abrasion, and ultraviolet radiation. Its disadvantages are that it is hard to determine damage to the core fibers, and it is more expensive than laid rope. Kernmantle rope is not readily identifiable as dynamic or static and must be marked accordingly.

c. Rope diameter and length vary with intended use. A standard diameter for many uses is 11 millimeters (7/16 inch). For stirrups, utility ropes, and hauling lines, 7 millimeters (5/16 inch), which can also be used in double strands, is adequate. Standards to remember are that the smaller the diameter, the less the tensile strength; but the larger the diameter, the heavier the rope. The standard lengths are 36 1/2 meters (120 feet), 40 meters (135 feet), 45 meters (150 feet), and 50 meters (165 feet). Normally, standard lengths should not be cut. If a nonstandard length is needed, the rope should be precut before the mission and clearly marked.

d. A rope should be selected for a specific use. There are different types of ropes for various uses. A dynamic rope is designed for climbing, while a static rope is designed for rappelling, rescue operations, load hauling, and rope installations.

WARNING

Using a rope for other than its intended use could result in personal injury or damage to equipment.

6-3. CARE OF ROPES

Because the rope is the rappeller's lifeline, it requires a great amount of care and maintenance.

a. Before and after use, the rope should be carefully inspected for excessive wear, cuts, mildew, or rotten spots. If any of these defects are found, the rope is unserviceable. An unserviceable rope is destroyed by cutting it into usable parts. If no sections are usable, the rope is cut into smaller parts and discarded.

b. Ropes should never be spliced together, stepped on, or dragged. This greatly reduces the tensile strength of the rope.

c. The rope must not come in contact with sharp edges. If this is likely, the areas must be padded or taped before rappelling.

d. The rope must be kept as dry as possible. If it becomes wet, it must be dried out to prevent rotting. (Avoid excessive heat/heating during the drying process.)

6-4. COILING

Two methods of coiling are the *mountaineer's coil* and the *farmer's coil* (Figure 6-3).

a. **Mountaineer's Coil.**

(1) To use the mountaineer's coil, grasp the rope about 3 feet from the end with the left hand. The right hand is then placed next to the left hand and runs along the rope until both arms are outstretched. Grasping the rope firmly, bring the hands together to form a loop, which is laid in the left hand. Repeat, forming uniform loops that run in a clockwise direction until the rope is completely coiled. If the rope tends to twist or form figure eights, give it a slight twist with the right hand when each loop is formed. Coil the rope in a clockwise direction to conform with the lay.

(2) To complete the coil, form a bight about 30 centimeters (12 inches) long with the starting end of the rope and lay it along the top of the coil. Uncoil the last loop and use it to wrap around the coil and the bight. Wrap toward the closed end of the bight, binding the final wrap across itself to lock it in place. Make six to eight wraps to secure the coil. Push the end of the rope down through the closed end of the bight. Carry the running end of the coil either in the pack (by forming a figure eight, doubling it, and placing it under the flap) or by placing it over the shoulder and under the opposite arm. If the rope to be coiled is anchored as in coiling a belay or rappel lane, start the coil near the end closest to the anchor so that the kinks work themselves out of the free end.

b. **Farmer's Coil.**

(1) The farmer's coil is easier and faster to coil than the mountaineer's coil. Establish the center of the rope either by locating its center mark or by grasping both ends of the rope and feeding them out until a bight comes up that is the center of the rope. Begin coiling with the doubled rope the same as the mountaineer's coil method. Leave about 4.6 meters (15 feet) of rope uncoiled. Then, squeeze the coils together and make four to six wraps around the middle of the coil, ensuring that the first wrap is held in place by the other wraps.

(2) Form a bight with the two running ends and place it through the bight formed by the top of the coil. Then, run the two running ends over the top of the coil and through the bight that is formed. Dress them in the center of the back of the carrier. Run the two ends over the shoulders to form shoulder straps. Then, bring them under the arms, cross them in the back over the coil, bring them around the body of the carrier, and tie them off with a square knot at the stomach.

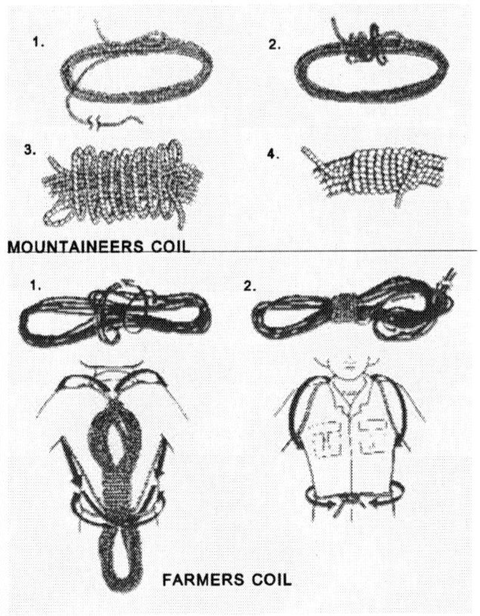

Figure 6-3. Two methods of tying a coil.

Section II. SNAPLINKS

Snaplinks are used to join equipment, rope, and people into a functioning system. When used properly, they are strong, versatile items. Snaplinks are metal (steel, aluminum, or alloys) loops with a hinged, spring-loaded gate on one side. They come in many sizes, shapes, and strengths (with and without locking gates) and in many types of metal. Steel is the strongest snaplink, but it is heavier.

6-5. DESCRIPTION

Snaplinks are available in many sizes and shapes to suit different needs. A locking snaplink is best for rappelling. Nonlocking snaplinks are easy to operate, but should be used only where they cannot be accidentally opened. Hollow snaplinks should be avoided since their use is limited. The following information applies to all snaplinks.

a. The weakest part of a snaplink is the gate. The gate must be closed before applying a load. When the gate is open, snaplinks should have little or no lateral movement of the gate.

b. Locking pins should be checked to ensure that they are not loose, worn, or corroded.

c. The metal should be checked for any cracks, grooves, burrs, flaws, or rust.

d. The spring-loaded gate should automatically close securely from an open to a closed position with no gap between the locking pin and notch.

e. If an engraver is used to mark snaplinks, it should be applied only to the gate, never to the load-bearing side.

f. A snaplink should never be side-loaded (across the gate) since this reduces the overall strength to the point of gate failure.

6-6. TYPES OF SNAPLINKS

The two types of snaplinks used during military operations are the standard snaplink (Figure 6-4) and the locking gate snaplink (Figure 6-5). Both snaplinks are available in a D-shaped variation or an oval variation.

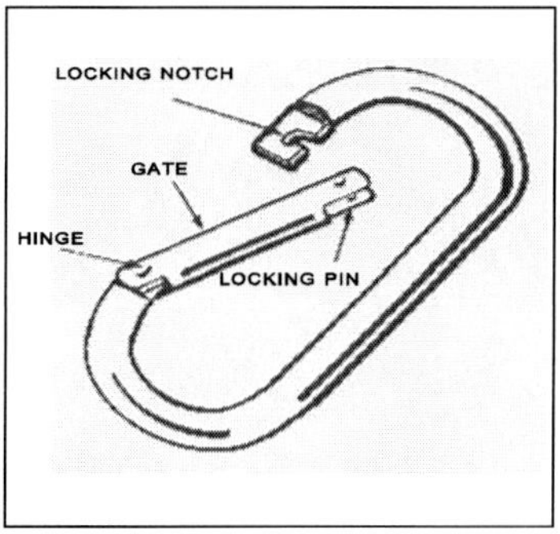

Figure 6-4. Standard snaplink (oval).

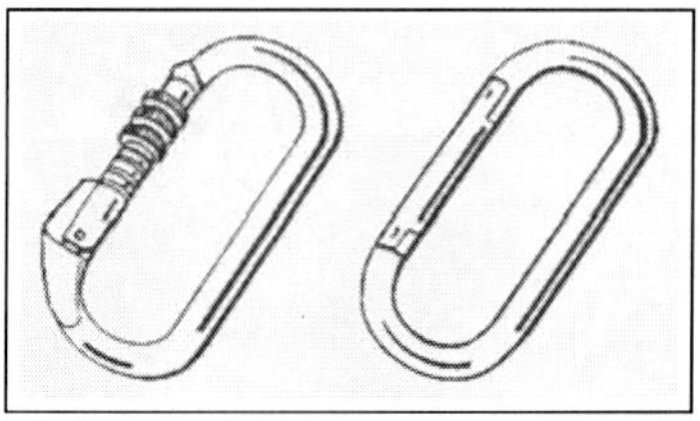

Figure 6-5. D-shaped snaplinks.

a. **D-Shaped Snaplinks.** D-shaped snaplinks are stronger than the oval type because the shape directs the largest part of the load to be applied to the longer, stronger side opposite the gate (Figure 6-5). D-shaped snaplinks are made of steel or aluminum alloys, and are available in many sizes and thicknesses, all of which can be with or without locking gates. Their strength and durability vary, which must be considered before use.

b. **Oval Snaplinks.** Oval snaplinks are versatile and have different applications. They are made of steel or aluminum alloys and are available in many sizes and thicknesses. Both sides of an oval snaplink bear the strain equally under load weight. Many modified ovals are available and most can be with or without the locking gate. All must meet military standards or UIAA specifications.

c. **Locking Snaplinks.** Locking snaplinks have a locking mechanism with a threaded sleeve on the gate. The sleeve screws tightly over the gate opening end or hinge end to hold the gate closed (Figure 6-6). A reverse locking gate is needed to prevent a moving rope from unscrewing the sleeve. The locking sleeve and threads should be kept free of dirt and grit. If the sleeve is forced to close, it may strip the threads. The locking mechanism ages and weakens after repeated use and should be routinely inspected for wear. The characteristics of locking snaplinks are as follows:

- Material: steel.
- Approximate strength: 2,000 pounds with the gate closed.
- Gate: spring-loaded without locking sleeve.
- Weight: 4.3 ounces.

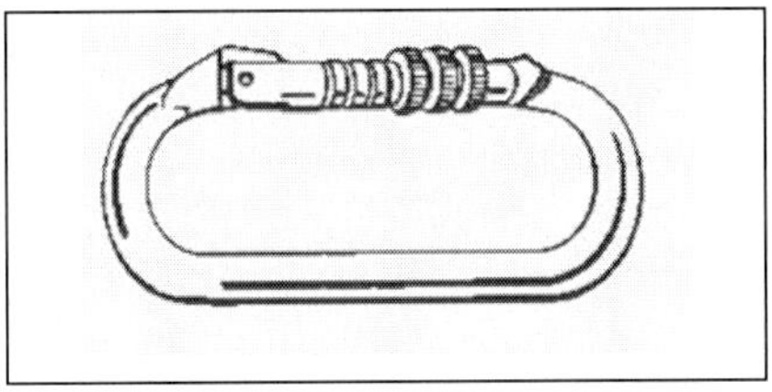

Figure 6-6. Locking snaplink.

6-7. INSPECTION

All rappelling material and individual rappelling equipment must be inspected before conducting a rappelling operation. The rappel master is responsible for ensuring the serviceability of rappelling material and individual rappelling equipment.

a. **Snaplinks.** Snaplinks should be inspected daily before, during, and after use.

(1) The metal should be checked for cracks, grooves, burrs, rust, and flaws. The gate should open and close freely without binding. There should be no lateral movement when the gate is open. The gate spring action should snap shut when released. The locking notch should have a slant or slot so that the gate remains shut under the impact of a rappeller's fall. The gate pins should not work their way out of their holes and should not be shorter than their holes. If there is a locking mechanism, it should be inspected to ensure that threads are not stripped and that the sleeve tightly locks the gate.

(2) If burrs, grooves, or rough areas are identified, the snaplink should not be used. Rust should be removed with steel wool and the spot rubbed with oil or solvent. The spring should be lubricated as needed. The snaplink must be boiled in water for 20 to 30 seconds to remove the cleaning agents since solvents and oils cause dirt to cling to the snaplink and to rub off on the ropes. It is better to use a dry graphite-based lubricant on snaplinks since such lubricants do not attract dirt.

b. **Ropes.** The ropes are inspected for serviceability. A rappel rope is unserviceable if it is saturated with petroleum products, is mildewed, shows excessive wear, has soft spots (internal damage), or if one strand is cut more than one-half of its diameter. Ropes should be inspected daily—before, during, and after use. An applicable annotation should be made in a corresponding rope log (DA Form 5752-R) (Figure 6-7).

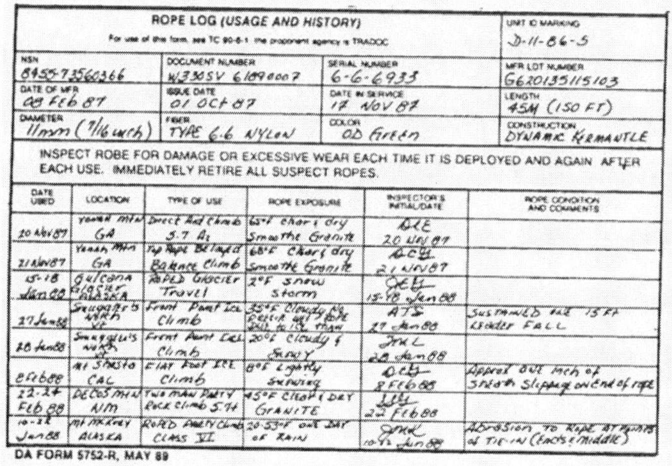

ROPE LOG (USAGE AND HISTORY)

For use of this form, see TC 90-6-1; the proponent agency is TRADOC

UNIT ID MARKING D-11-86-5			
NSN 8455-7356036	DOCUMENT NUMBER W330SV 61890007	SERIAL NUMBER 6-6-6935	MFR LOT NUMBER G6201351151O3
DATE OF MFR 08 Feb 87	ISSUE DATE 01 Oct 87	DATE IN SERVICE 17 NOV 87	LENGTH 45M (150 FT)
DIAMETER 11mm (7/16 inch)	FIBER TYPE 6.6 NYLON	COLOR OD Green	CONSTRUCTION DYNAMIC KERMANTLE

INSPECT ROBE FOR DAMAGE OR EXCESSIVE WEAR EACH TIME IT IS DEPLOYED AND AGAIN AFTER EACH USE. IMMEDIATELY RETIRE ALL SUSPECT ROPES.

DATE USED	LOCATION	TYPE OF USE	ROPE EXPOSURE	INSPECTOR'S INITIAL/DATE	ROPE CONDITION AND COMMENTS
20 Nov 87	Yonah Mtn GA	Direct Aid Climb 5.7 A1	65°F Clear & dry Smooth Granite	[illegible] 20 Nov 87	
21 Nov 87	Yonah Mtn GA	Top Rope Belayed Balance Climb	68°F Clear & dry Smooth Granite	[illegible] 21 Nov 87	
15-18 Jan 88	Gulcana Glacier ALASKA	Roped Glacier Travel	2°F Snow Storm	[illegible] 15-18 Jan 88	
27 Jan 88	Smugglers Notch VT	Front Point Ice Climb	35°F Cloudy [illegible]	ATS 27 Jan 88	Sustained 15 ft Leader Fall
28 Jan 88	Smugglers Notch VT	Front Point Ice Climb	20°F Cloudy & Snowy	[illegible] 28 Jan 88	
8 Feb 88	Mt Shasta CAL	Flat Foot Ice Climb	8°F Lightly Snowing	[illegible] 8 Feb 88	Approx one inch of Sheath Slipping on end of rope
22-24 Feb 88	DeCos Mtn NM	Two Man Party Rock Climb 5.7+	45°F Clear & Dry Granite	[illegible] 22 Feb 88	
10-22 Jun 88	Mt McKinley ALASKA	Roped Party Climb Class VI	20-53°F one day of Rain	[illegible] 10-22 Jun 88	Abrasion to Rope at points of Tie-in (Ends & middle)

DA FORM 5752-R, MAY 89

Figure 6-7. Example of completed DA Form 5752-R.

c. **Gloves.** The gloves are inspected for serviceability. They are unserviceable if holes are found in the friction-bearing surfaces or along the seams.

d. **Helmet.** The helmet is inspected for correct assembly and serviceability. It is unserviceable if it cannot be properly assembled or if any straps are cut or torn one-quarter of its width.

(1) Ensure that the chin strap attaching tabs are secured to the helmet by the locking nuts.

(2) Ensure that the pull-the-dot tab has four plies of nylon and that the chin strap is routed from the inside of the buckle around the horizontal bar and back toward the inside of the buckle.

Section III. ALTERNATE METHODS OF DESCENT

Snaplinks or carabiners that are used for rappelling are placed through the sewn loops of the seat harness or the sewn loops of the seat-chest combination.

6-8. FIGURE-EIGHT DESCENDER

The rappeller attaches a locking carabiner to the harness (Figure 6-8). He routes the rappel rope up through the large hole and places the bight over the collar. He inserts the figure eight with the rappel rope attached into the locking carabiner, and locks it down. The rappeller brakes to the rear and descends as in a seat hip rappel. If heavy loads are to be descended, he uses a double wrap around the collar.

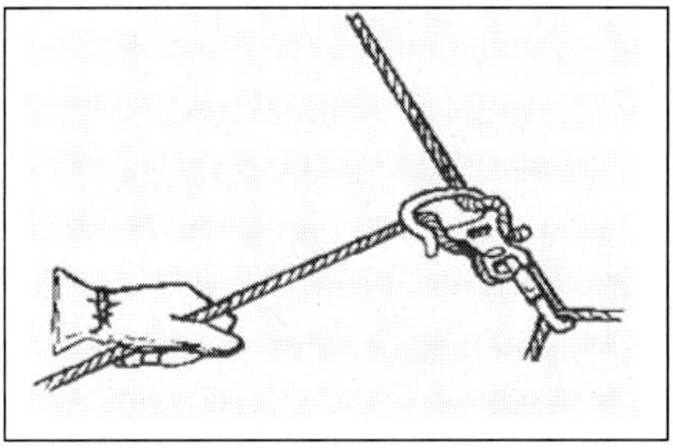

Figure 6-8. Figure-eight descender.

6-9. MUNTER HITCH

The rappeller attaches a large radius snaplink to the harness or rappel seat. He ties a Munter hitch and clip into the snaplink (Figure 6-9, page 6-9). He brakes to the front and descends with an L-shape body position.

Notes: 1. The guide hand should remain on the standing end of the rope to keep the rappeller's head and face away from the hardware.

2. The Munter hitch creates significant rope-to-rope friction and may cause premature wear on nylon ropes.

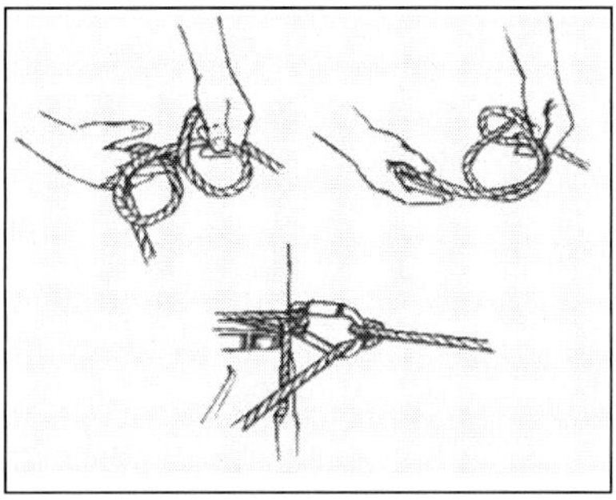

Figure 6-9. Munter hitch.

CHAPTER 7
KNOTS

A knot is a fastening made by intertwining or tying pieces of string, cord, rope, or webbing. All knots used by a rappeller are divided into four classes: Class I—joining knots, Class II—anchor knots, Class III—middle rope knots, and Class IV—special knots. These classes of knots are intended only as a general guide since some of the knots discussed herein may be appropriate in more than one class. The skill of knot tying may be lost if it is not used and practiced. With experience and practice, knot tying becomes instinctive and helps the soldier in many situations.

7-1. MILITARY MOUNTAINEERING TERMINOLOGY

The terms explained below are the most commonly used in military mountaineering (Figure 7-1).

a. A *bight of rope* is a simple bend of rope in which the rope does not cross itself.

b. A *loop* is a bend of rope in which the rope does cross itself.

c. A *half hitch* is a loop that runs around an object and locks itself.

d. The *running end* (working end) of the rope is the free end of the rope that can be used.

e. The *standing part* of the rope is the part that is static (anchored, coiled); the remaining part of the rope not being used (also called static end).

f. The *lay* of the rope is the same as the twist of the rope.

g. A *round turn* is a single complete wrap of the rope around an object providing 360-degree contact. The running end leaves the completed circle in the same direction as the standing part. In a round turn, the rope is wrapped around an object 1 1/2 times.

h. The *pigtail* is the short length of rope remaining at the end after tying a knot or coiling a rope.

i. *Back feeding* (or stacking) is taking off one wrap at a time from a coil, and letting it fall naturally to the ground.

j. *Dress down the knot* is tightening down a knot to its functioning form.

k. An *overhand* is a loop with the running end pulled through the loop.

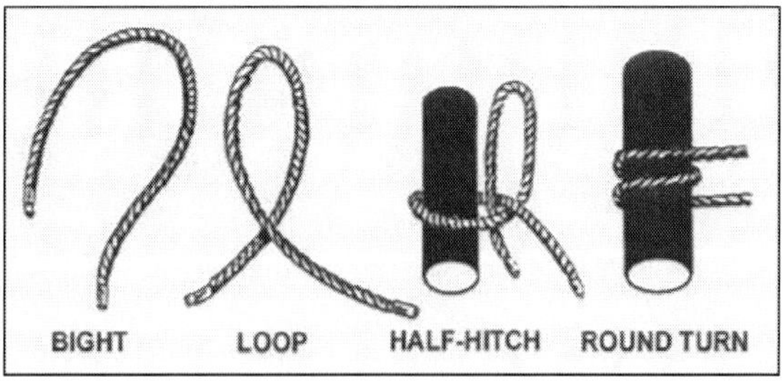

Figure 7-1. Examples of rope terminology.

Section I. JOINING KNOTS

7-2. SQUARE KNOT

The square knot is used to tie the ends of two ropes of equal diameter (Figure 7-2).

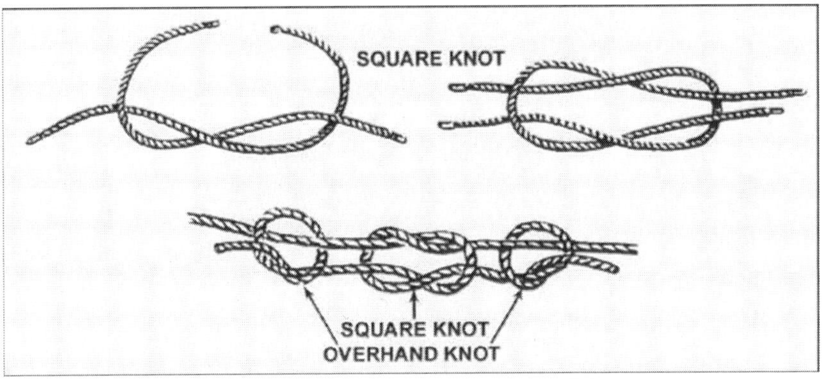

Figure 7-2. Square knot.

a. **Tying the Knot.**

STEP 1. Holding one working end in each hand, place the working end in the right hand over the one in the left hand.
STEP 2. Pull it under and back over the top of the rope in the left hand.
STEP 3. Place the working end in the left hand over the one in the right hand and repeat STEP 2.
STEP 4. Dress the knot down and secure it with an overhand knot on each side of the square knot.

b. **Checkpoints.**

(1) There are two interlocking bights.

(2) The standing parts are on the same side and properly secured with overhand knots.

7-3. DOUBLE SHEET BEND KNOT

The double sheet bend knot is used to tie the ends of two or more ropes of equal or unequal diameter (Figure 7-3). When a single rope is tied to a multiple of ropes, the bight is formed with the multiple of ropes.

Figure 7-3. Double sheet bend knot.

a. **Tying the Knot.**

STEP 1. Form a 30-centimeter (12-inch) bight in the left hand with the rope or ropes. Ensure that the short end of the bight is facing to the inside.

STEP 2. Place the index finger of the left hand on top of the bight. Bring the rope in the right hand up through the bight and over the index finger of the left hand, so the working end is away from the body.

STEP 3. Reach through the loop and grasp the working end of the rope and bring it back towards the body. Place it between the index finger and the bight, forming a round turn. Repeat this one more time, wrapping toward the tip of the index finger.

STEP 4. Remove the finger and dress the knot down.

b. **Checkpoints.**

(1) The two wraps around the bight are held in place by a locking bar.

(2) The two standing parts of the ropes exit parallel and from opposite ends of the knot. The two working ends form an L and exit the knot at a 90-degree angle and on opposite sides of the knot.

7-4. FISHERMAN'S KNOT

The fisherman's knot is used to tie two ropes of similar or dissimilar materials (Figure 7-4).

Figure 7-4. Fisherman's knot.

a. **Tying the Knot.**

STEP 1. Tie an overhand knot in one end of the rope.
STEP 2. Pass the working end of the other rope through the first overhand knot. Tie an overhand knot around the standing part of the first rope with the working end of the second rope.
STEP 3. Tightly dress down each overhand knot and tightly draw the knots together.

b. **Checkpoints.**

(1) The two separate overhand knots are tied tightly around the long standing part of the opposing rope.

(2) The two overhand knots are drawn snug.

7-5. DOUBLE FISHERMAN'S KNOT

The double fisherman's knot (also called double English or grapevine) is used to tie two ropes of similar or dissimilar materials (Figure 7-5).

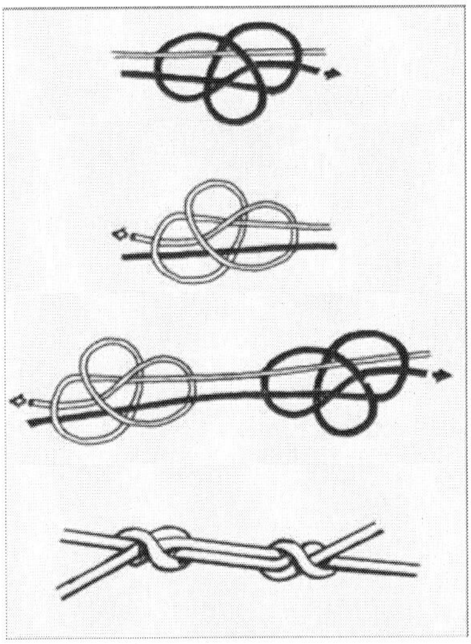

Figure 7-5. Double fisherman's knot.

a. **Tying the Knot.**

STEP 1. With the working end of one rope, tie two wraps around the standing part of another rope.
STEP 2. Insert the working end (STEP 1) back through the two wraps and draw it tight.

STEP 3. With the working end of the other rope, which contains the standing part (STEPS 1 and 2), tie two wraps around the standing part of the other rope (the working end in STEP 1). Insert the working end back through the two wraps and draw tight.

STEP 4. Pull on the opposing ends to bring the two knots together.

b. **Checkpoints.**

(1) Two wraps with the working end running through and drawn tight.

(2) Both wraps are drawn together with all the slack out of the knot.

7-6. WATER KNOT

The water knot is used to attach two webbing ends (Figure 7-6). It is also called a ring bend, overhand retrace, or tape knot. It is used in runners and harnesses.

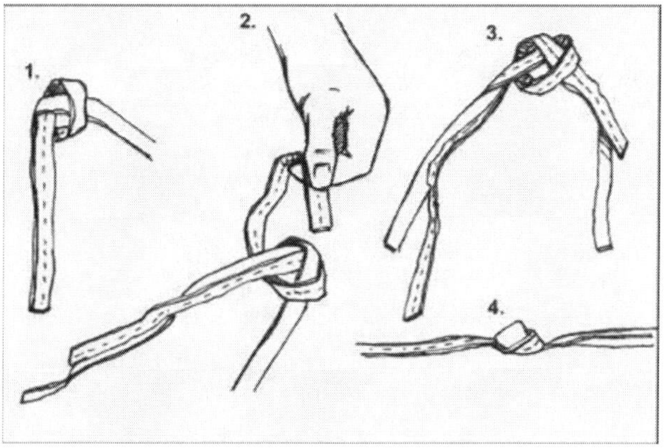

Figure 7-6. Water knot

a. **Tying the Knot.**

STEP 1. Tie an overhand knot in one of the ends.

STEP 2. Feed the other end back through the knot, following the path of the first rope in reverse.

b. **Checkpoints.**

(1) There are two overhand knots, one retracing the other.

(2) There is no slack in the knot, with the working ends coming out of the knot in opposite directions (at least 5 centimeters [2 inches] long).

Section II. ANCHOR KNOTS

7-7. BOWLINE KNOT

The bowline knot is used to tie a single fixed loop in the end of a rope (Figure 7-7). It does not slip under strain and is easily untied. This knot is always used when there is alternating tension. It can also be used to tie the end of a rope to an anchor.

Figure 7-7. Bowline knot.

a. **Tying the Knot.**

STEP 1. Bring the working end of the rope around the anchor, from right to left (as the rappeller faces the anchor).

STEP 2. Form an overhand loop in the standing part of the rope (on the rappeller's right) toward the anchor.

STEP 3. Reach through the loop and pull up a bight.

STEP 4. Place the working end of the rope (on the rappeller's left) through the bight, and bring it back onto itself. Dress the knot down.

STEP 5. Form an overhand knot with the tail from the bight.

b. **Checkpoints.**

(1) The loop is locked into place by a bight.

(2) The short portion of the bight is on the inside and on the loop around the anchor (or inside the fixed loop).

7-8. ROUND TURN AND TWO HALF HITCHES KNOT The round turn and two half hitches knot is used to tie the end of a rope to an anchor and must have constant tension (Figure 7-8).

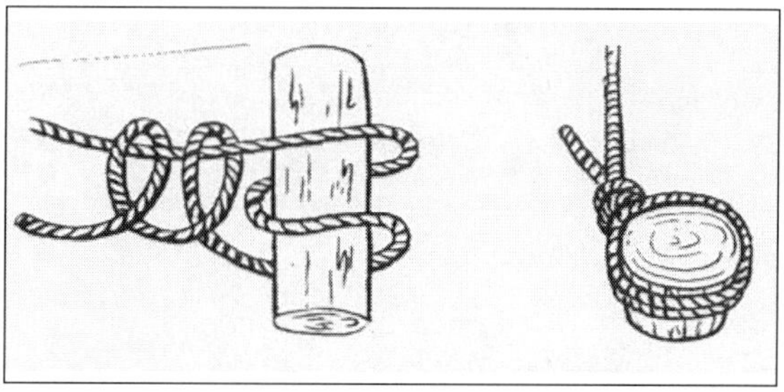

Figure 7-8. Round turn and two half hitches knot.

a. **Tying the Knot.**

STEP 1. Route the rope around the anchor from right to left and wrap down (must have two wraps in the rear of the anchor and one in the front). Run the loop around the object to provide 360-degree contact, distributing the load over the anchor.

STEP 2. Bring the working end of the rope left to right and over the standing part, forming a half hitch (first half hitch).

STEP 3. Repeat STEP 2 (last half hitch has a 15-centimeter [6-inch] pigtail).

STEP 4. Dress the knot down.

b. **Checkpoints.**

(1) A complete round turn should exist around the anchor with no crosses.

(2) Two half hitches should be held in place by a locking bar with no less than a 15-centimeter (6-inch) tail remaining.

(3) When the two half hitches are twisted to the left or the right, they will appear to for a figure eight.

(4) Looking at the top of the half hitches, the rope should appear to be one continuous rope.

7-9. FIGURE-EIGHT RETRACE KNOT

The figure-eight retrace knot is also called a rerouted figure eight. Although it produces the same result as a figure-eight loop, by tying the knot in a retrace it can be used to fasten the rope to trees or to places where the loop cannot be used (Figure 7-9).

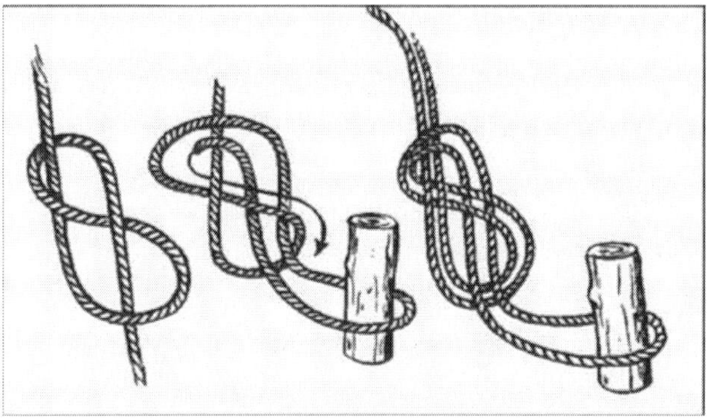

Figure 7-9. Figure-eight retrace.

a. **Tying the Knot.**

STEP 1. Use a length of rope long enough to go around the anchor, leaving enough rope to work with.
STEP 2. Tie a figure-eight knot in the standing part of the rope, leaving enough rope to go around the anchor. To tie a figure-eight knot:

- Form a loop in the rope.
- Wrap the working end around the standing part.
- Route the working end through the loop.
- The finished knot is dressed loosely.

STEP 3. Place the working end around the anchor point.
STEP 4. With the working end, insert the rope back through the loop of the knot in reverse.
STEP 5. Keep the original figure eight as the outside rope and retrace the knot around the wrap and back to the long sttanding part.
STEP 6. Safety off with an overhand knot.

b. **Checkpoints.**

(1) At least 15 centimeters (6 inches) of rope extends past the figure-eight knot.

(2) The original figure eight, tied with the standing end, remains as the outermost rope throughout the knot.

(3) The knot is dressed tightly.

7-10. CLOVE HITCH KNOT

The clove hitch knot can be used in the middle of the rope as well as at the end of the rope (Figure 7-10). The knot must have constant tension on it once tied to prevent slipping. It can be used as either an anchor or middle knot, depending on how it is tied. (Do not use a clove hitch knot on metal except on a picket-hold-fast.)

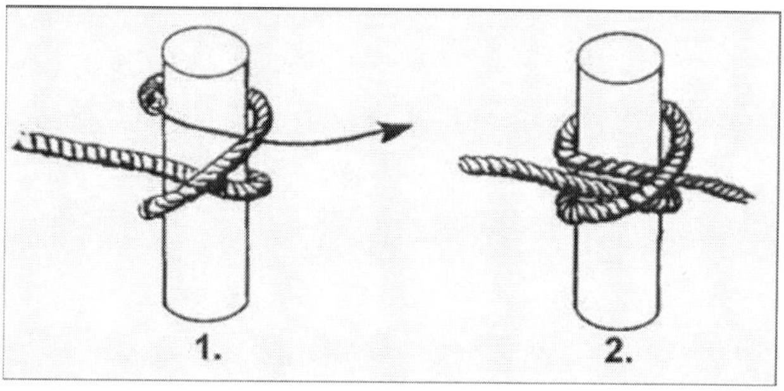

Figure 7-10. Clove hitch knot.

a. **Tying the Knot.**

(1) If there is access over the top of the anchor, the knot is tied as follows:

STEP 1. Hold rope in both hands, palms down with hands together. Slide the left hand to the left from 20 to 25 centimeters (8 to 10 inches).

STEP 2. Form a loop away from and back toward the right.

STEP 3. Slide the right hand from 20 to 25 centimeters (8 to 10 inches) to the right. Form a loop inward and back to the left hand.

STEP 4. Place the left loop on top of the right loop. Place both loops over the anchor and pull both ends of the rope in opposite directions to complete the knot.

(2) If the clove hitch knot must go around the anchor, tie the knot as follows (for instructional purposes, assume that the anchor is horizontal):

STEP 1. Place 76 centimeters (30 inches) of rope over the top of the anchor. Hold the standing end in the left hand. With the right hand, reach under the horizontal anchor, grasp the working end, and bring it inward.

STEP 2. Place the working end of the rope over the standing end (to form a loop). Hold the loop in the left hand. Place the working end over the anchor from 20 to 25 centimeters (8 to 10 inches) to the left of the loop.

STEP 3. With the right hand, reach down to the left-hand side of the loop under the anchor. Grasp the working end of the rope and bring it up and outward.

STEP 4. Dress down the knot.

b. **Checkpoints.**
(1) The knot has two round turns around the anchor with locking bar.
(2) The locking bar is facing 90 degrees from the direction of pull.
(3) The ends exit 180 degrees from each other.
(4) The knot has more than a 15-centimeter (6-inch) pigtail remaining.

SECTION III. MIDDLE ROPE KNOTS

7-11. WIREMAN'S KNOT

The wireman's knot forms a single, fixed loop in the middle of the rope (Figure 7-11).

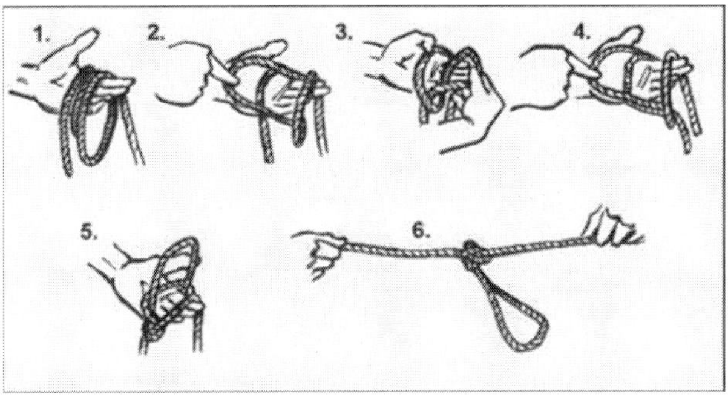

Figure 7-11. Wireman's knot.

a. **Tying the Knot.**

STEP 1. When tying this knot, face the anchor that the tie-off system will be tied to. Take up the slack from the anchor, and wrap two turns around the left hand (palm up) from left to right.
STEP 2. Take up a loop of 30 centimeters (12 inches) in the second round turn to create the fixed loop of the knot.
STEP 3. Name the wraps from the palm to the fingertips: heel, palm, and fingertip.
STEP 4. Grasp the palm wrap with the right thumb and forefinger, and place it over the heel wrap.
STEP 5. Grasp the heel wrap and place it over the fingertip wrap.
STEP 6. Grasp the fingertip wrap and place it over the palm wrap.
STEP 7. Grasp the palm wrap and pull up to form a fixed loop.
STEP 8. Dress the knot down by pulling on the fixed loop and the two working ends.
STEP 9. Pull the working ends apart to finish the knot.

b. **Checkpoints.**

(1) The completed knot should have four separate bights locking down on themselves with the fixed loop exiting from the top of the knot and laying toward the near side anchor point.

(2) Both ends should exit opposite each other without any bends.

7-12. DIRECTIONAL FIGURE-EIGHT KNOT

The directional figure-eight knot forms a single, fixed loop in the middle of the rope that lays back along the standing part of the rope (Figure 7-12).

Figure 7-12. Directional figure-eight knot.

a. **Tying the Knot.**

STEP 1. Face the far side anchor so that when the knot is tied, it lays inward.
STEP 2. Lay the rope from the far side anchor over the left palm. Make one wrap around the palm.
STEP 3. With the wrap thus formed, tie a figure-eight knot around the standing part that leads to the far side anchor.
STEP 4. Dress the knot down with the tail and the bight together.

b. **Checkpoints.**

(1) The loop should be large enough to accept a snaplink but no larger than a helmet-size loop.

(2) The tail and bight must be together.

(3) The figure eight is tied tightly.

(4) The bight in the knot faces back toward the near side.

7-13. BOWLINE-ON-A-BIGHT KNOT

The bowline-on-a-bight knot is used to form two fixed loops in the middle of a rope (Figure 7-13).

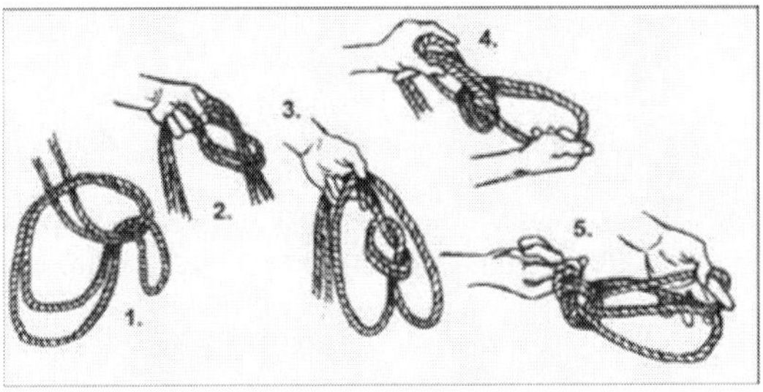

Figure 7-13. Bowline-on-a-bight knot.

a. **Tying the Knot.**

STEP 1. Form a bight in the rope about twice as long as the finished loops will be.
STEP 2. Tie an overhand knot on a bight.
STEP 3. Hold the overhand knot in the left hand so that the bight is running down and outward.
STEP 4. Grasp the bight with the right hand, fold it back over the overhand knot so that the overhand knot goes through the bight.
STEP 5. From the end (apex) of the bight, follow the bight back to where it forms the cross in the overhand knot. Grasp the two ropes that run down and outward and pull up, forming two loops.
STEP 6. Pull the two ropes out of the overhand knot and dress the knot down.
STEP 7. A final dress is required: grasp the ends of the two fixed loops and pull, spreading them apart to ensure the loops do not slip.

b. **Checkpoints.**

(1) There are two fixed loops that will not slip.
(2) There are no twists in the knot.
(3) A double loop is held in place by a bight.

7-14. FIGURE-EIGHT-ON-A-BIGHT KNOT

The figure-eight-on-a-bight knot is used to form two fixed loops in the middle of a rope . (Figure 7-14).

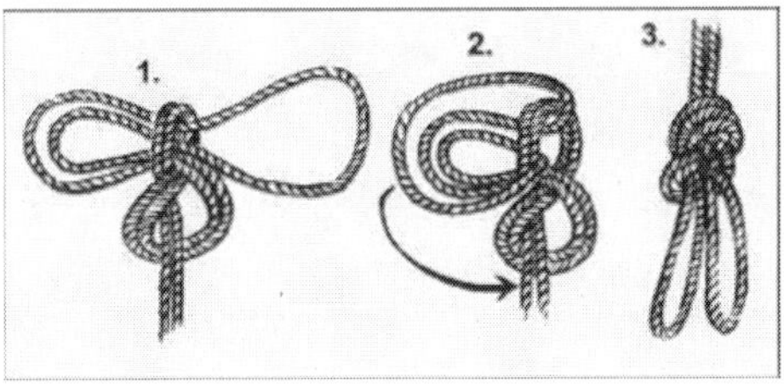

Figure 7-14. Figure-eight-on-a-bight knot.

a. **Tying the Knot.**

STEP 1. Using the doubled rope, form a 46-centimeter (18-inch) bight in the left hand with the running end facing to the left.
STEP 2. Grasp the bight with the right hand and make a 360-degree turn around the standing end in a counterclockwise direction.
STEP 3. With the working end, form another bight and place that bight through the loop just formed in the left hand.
STEP 4. Hold the bight with the left hand, and place the original bight (moving toward the left hand) over the knot.
STEP 5. Dress the knot down.

b. **Checkpoints.**
(1) There is a double figure eight with two loops.
(2) The knot will not slip.

7-15. OVERHAND LOOP KNOT

The overhand loop knot forms a single loop in the middle of a rope. It should not be used in a transport tightening system (Figure 7-15).

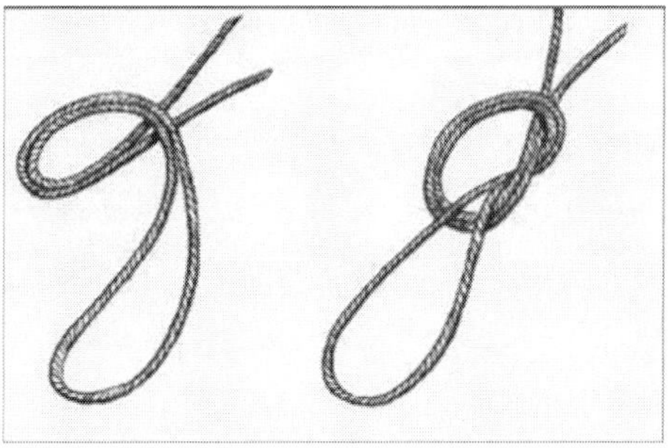

Figure 7-15. Overhand loop knot

a. **Tying the Knot.**
STEP 1. Form a bight in the rope about as long as the loop needed.
STEP 2. With the bight, form an overhand knot using the bight as the working end. Dress the knot tightly.

b. **Checkpoints.**
(1) The loop is the desired size.
(2) The ropes in the overhand knot are parallel and do not cross over each other.
(3) The knot is tightly dressed.

7-16. FIGURE-EIGHT LOOP KNOT

The figure-eight loop knot is a variation of the overhand loop and is easy to untie after it has held weight (Figure 7-16).

Figure 7-16. Figure eight loop knot.

a. **Tying the Knot.**

STEP 1. Form a bight in the rope about as large as the diameter of the desired loop.
STEP 2. With the bight as the working end, form a loop in rope (standing part).
STEP 3. Wrap the working end around the standing part 360 degrees and feed the working end through the loop. Dress the knot tightly.

b. **Checkpoints.**

(1) The loop is the desired size.
(2) The ropes in the loop are parallel and do not cross over each other.
(3) The knot is tightly dressed.

SECTION IV. SPECIAL KNOTS

7-17. SINGLE BUTTERFLY KNOT

The single butterfly knot is used to form a single, fixed loop in the middle of the rope without using the ends (Figure 7-17). The butterfly can be used for the middle man in a rope party as well as in a transport tightening system. The knot can be hard to untie when heavy weight has been placed on it for extended periods.

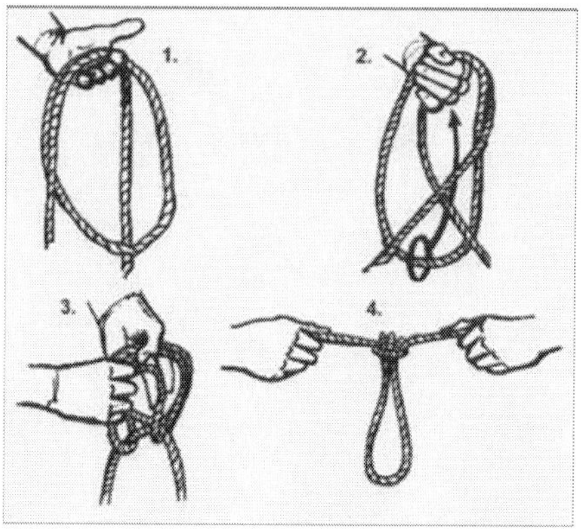

Figure 7-17. Single butterfly knot.

a. **Tying the Knot.**

STEP 1. Form a loop over the palm of the left hand.
STEP 2. With the right hand, cross the two long-standing ends; reach under and grasp the bottom of the loop formed in STEP 1.
STEP 3. Route the bottom of the loop over the crossed standing ends and back through the top of the loop.
STEP 4. Hold the newly formed loop and dress the knot down, removing all of the slack out of the knot.

b. **Checkpoints.**
(1) The wings of the knot are dressed down tightly and as close together as possible.
(2) The ropes between the wings are parallel with no crossovers.
(3) The loop is no larger than a helmet but large enough to accept a snaplink.
(4) All ropes in the knot are tightly dressed.

7-18. PRUSIK KNOT

The prusik knot is used to put a moveable rope on a fixed rope such as a prusik ascent or a tightening system. This knot can be tied as a middle, finger, or end knot.

a. **Tying the Knot.**

(1) *First method.* The prusik knot is tied at the middle or end of a rope; there are many ways to do this (Figure 7-18). The middle-of-the-rope prusik knot can be tied with a short rope to a long rope as follows:

STEP 1. Double the short rope, forming a bight, with the working ends even. Lay it over the long rope so that the closed end of the bight is 30 centimeters (12 inches) below the long rope and the remaining part of the rope (working ends) is the closest to the rappeller; spread the working end apart.

STEP 2. Reach down through the 30-centimeter (12-inch) bight. Pull up both of the working ends and lay them over the long rope. Repeat this process making sure that the working ends pass in the middle of the first two wraps. Now there are four wraps and a locking bar working across them on the long rope.

STEP 3. Dress the wraps and locking bar down to ensure they are tight and not twisted. Tie an overhand knot with both ropes to prevent the knot from slipping during periods of variable tension.

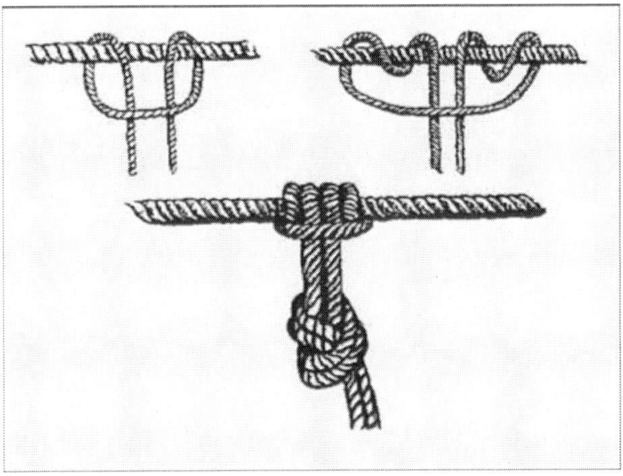

Figure 7-18. Middle-of-the-rope prusik knot.

(2) *Second method.* Another way of tying the prusik knot is the finger prusik (Figure 7-19).

STEP 1. Lay the rope across the thumb and index finger of the left hand.

STEP 2. Wrap the rope around the index finger twice (wrapping away from the rappeller). Wrap the rope around the thumb twice (wrapping towards the rappeller).

STEP 3. Place the thumb and index finger tip to tip. Push the wraps together so that the prusik is on the left thumb. Four wraps and a locking bar should be across them.

STEP 4. Pass the running end of the rope through the wraps of the prusik knot.

STEP 5. Dress the wraps and locking bar down to ensure they are tight and not twisted.

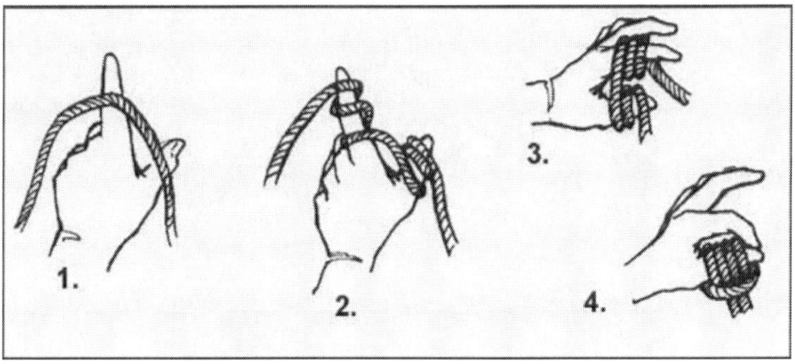

Figure 7-19. Finger prusik knot.

(3) *Third method.* Another way of tying a prusik knot is the end-of-the-rope knot (Figure 7-20).

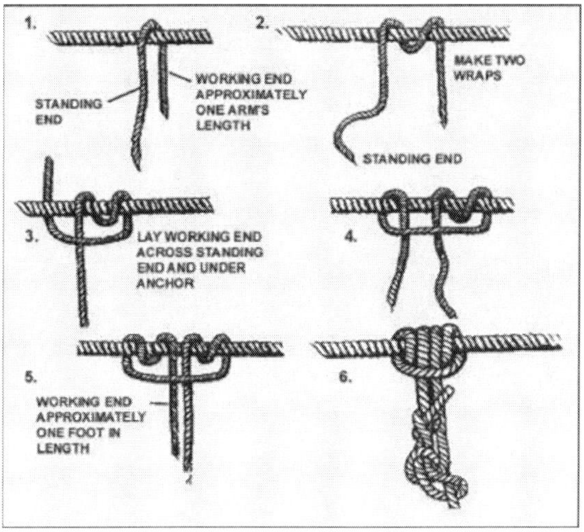

Figure 7-20. End-of-the-rope prusik knot.

STEP 1. Using an arm's length of rope, place it over the long rope.
STEP 2. Form a complete round turn in the rope.
STEP 3. Cross over the standing part of the short rope with the working end of the short rope.
STEP 4. Lay the working end under the long rope.
STEP 5. Form a complete round turn in the rope, working back toward the middle of the knot.

STEP 6. Four wraps and a locking bar should run across them on the long rope. Dress the wraps and locking bar down. Ensure they are tight, parallel, and not twisted.

STEP 7. Finish the knot with a bowline to ensure that the prusik knot will not slip out during periods of varying tension.

b. **Checkpoints.**

(1) Four wraps with a locking bar.

(2) The locking bar faces the rappeller.

(3) The knot is tight and dressed down with no ropes twisted or crossed.

(4) Other than a finger prusik, the knot should contain an overhand or bowline to prevent slipping.

7-19. BACHMAN KNOT

The bachman knot provides a means of using a makeshift mechanized ascender (Figure 7-21).

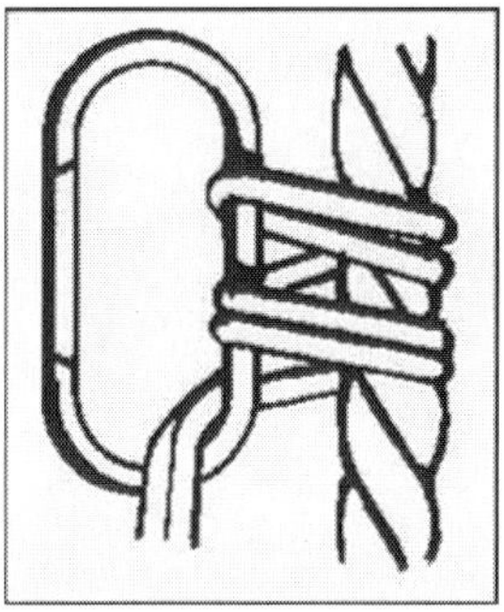

Figure 7-21. Bachman knot.

a. **Tying the Knot.**

STEP 1. Find the middle of a utility rope and insert it into a snaplink.

STEP 2. Place the snaplink and utility rope next to a long climbing rope.

STEP 3. With the two ropes parallel fromthe snaplink, make two or more wraps wraps around the climbing rope and through the inside portion of the snaplink.

Note: The rope can be tied into an etrier (stirrup) and used as a prusik-friction principal ascender.

b. **Checkpoints.**

(1) The bight of the climbing rope is at the top of the snaplink.

(2) The two ropes run parallel without twisting or crossing.

(3) Two or more wraps are made around the long climbing rope and through the inside portion of the snaplink.

7-20. BOWLINE-ON-A-COIL KNOT

The bowline-on-a-coil knot is an expedient tie-in used by rappellers in two- and three-man party climbs when a climbing harness is not available (Figure 7-22).

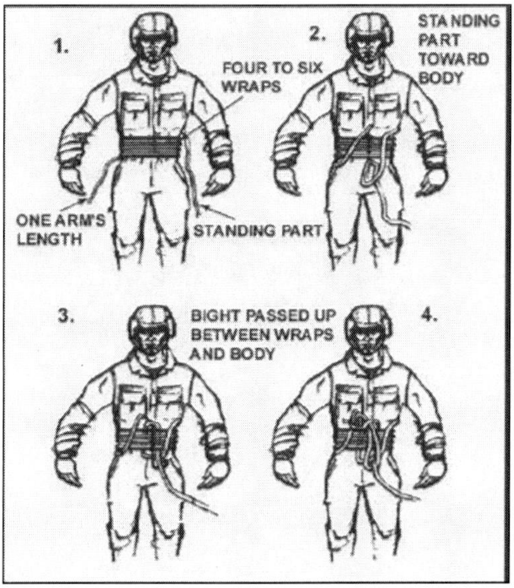

Figure 7-22. Bowline-on-a-coil knot.

Note: A safer tie in, when possible, is a swiss seat and figure eight around the shoulder, to avoid damaging the ribs if the climber falls.

a. **Tying the Knot.**

STEP 1. To tie a bowline-on-a-coil knot, start the same as tying a bowline around the waist. (The waist of the rappeller is the anchor point. The rappeller has about 40 centimeters [16 inches] of rope in the right hand. The standing end is coming from the remainder of the rope in his left hand.)

STEP 2. Draw slack from the standing end of the rope in the left hand, wrapping enough coils around the body to complete a minimum of four (but no more than six) coils.

STEP 3. With the rope in the left hand, make a loop inward with the standing end.

STEP 4. Slide the loop up between the horizontal wraps and the body, ensuring that the loop does not come uncrossed.

STEP 5. Bring a bight up through the loop. Bring the working end of the rope in the right hand through the bight and back onto itself.

OR

Insert the short working end to the right through the loop (just passed through the wraps). Pass the working end around the long standing end and back on itself, forming a bight.

STEP 6. Dress the knot down.

STEP 7. Safety off with an overhand knot against the knot formed on the top single coil. From 10 to 15 centimeters (4 to 6 inches) of rope should be left (standing part).

b. **Checkpoints.**

(1) There are a minimum of four (maximum of six) parallel wraps; the top and bottom ropes cross forward of the hips.

(2) The loop must be underneath all wraps.

(3) Other checkpoints are the same as the bowline.

7-21. THREE-LOOP BOWLINE KNOT

The three-loop bowline knot is used to form three fixed loops in the middle of a rope (Figure 7-23). It is used in a self-equalizing anchor system.

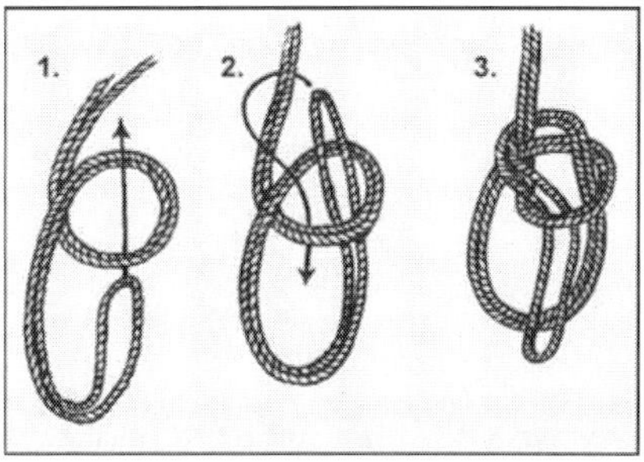

Figure 7-23. Three-loop bowline knot.

a. **Tying the Knot.**

STEP 1. Form about a 60 centimeter (24-inch) bight in the rope, laying it across the left hand, palm up, with the bight to the front.

STEP 2. With the right hand, form a loop with both ropes. Hold it in the left hand.

STEP 3. With the right hand, bring the long bight through the loop(formed in STEP 2) and around and around the two standing ends.

STEP 4. Follow the bight around the standing ends and back down into the loop.

STEP 5. Hold all three loops in the left hand. With the right hand, grasp both standing ropes and pull while dressing up the bowline.

b. **Checkpoints.**

(1) Checkpoints are the same as a regular bowline, except that it is tied with two ropes.

(2) An overhand knot must be tied if the third loop is not to be used as a secondary anchor.

7-22. FIGURE-EIGHT SLIP KNOT

The figure-eight slip knot forms an adjustable bight in a rope (Figure 7-24).

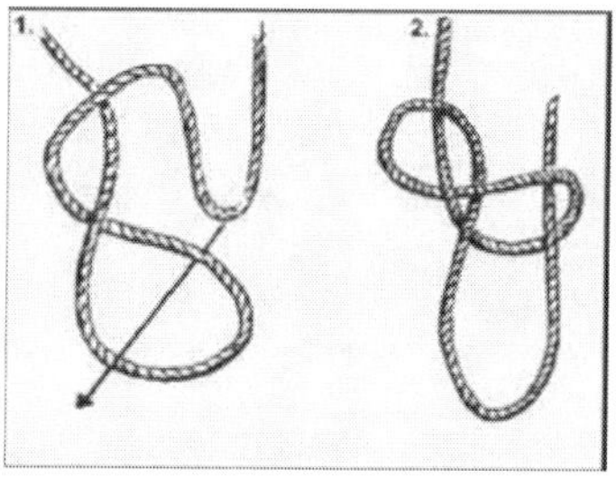

Figure 7-24. Figure-eight slip knot.

a. **Tying the Knot.**

STEP 1. Form a 30-centimeter (12-inch) bight in the end of the rope.

STEP 2. Hold the center of the bight in the right hand. Hold the two parallel ropes from the bight in the left hand about 30 centimeters (12 inches) up the rope.

STEP 3. With the center of the bight in the right hand, twist two complete turns clockwise.

STEP 4. Reach through the bight and grasp the long standing end of the rope. Pull another bight (from the long standing end) back through the original bight.

STEP 5. Pull down on the short working end of the rope and dress the knot down.

STEP 6. If the knot is to be used in a transport tightening system, take the working end of the rope and form a half hitch around the loop of the figure-eight knot.

b. **Checkpoints.**

(1) The knot is in the shape of a figure eight.

(2) The sliding portion of the rope is the long working end of the rope.

Section V. SPECIAL-PURPOSE KNOTS

7-23. KLEIMHIEST KNOT

The kleimhiest knot provides a moveable, easily adjustable, high-tension knot capable of holding extremely heavy loads while being pulled tight (Figure 7-25). This system requires a six-man pulling team when used in a transport tightening system.

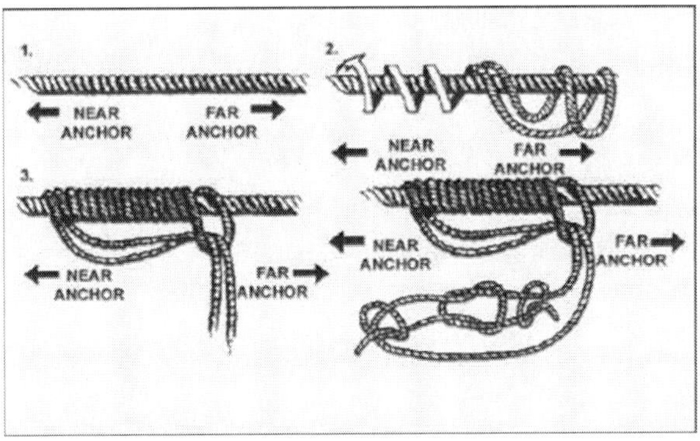

Figure 7-25. Kleimhiest knot.

a. **Tying the Knot.**

STEP 1. Using a utility rope, offset the ends by 30 centimeters (12inches). With the ends offset, find the center of the rope and form a bight. Lay the bight over a horizontal rope near the far side anchor.

STEP 2. Wrap the tails of the utility rope around the horizontal rope back toward the near side anchor. Wrap at least four complete turns.

STEP 3. Pass the remaining tails of the utility rope through the bight (see STEP 1).

STEP 4. Join the two ends of the tail with a joining knot (square knot with overhand safety knots).

STEP 5. Dress the knot down tightly so that all wraps are touching. Insert two snaplinks (opposing gates) into the large loop tied with the tails.

b. **Checkpoints.**

(1) The bight is closest to the far side anchor.

(2) All wraps are tight and touching.

(3) The ends of the utility rope are properly secured with the adjoining knot.

7-24. OVERHAND KNOT

The overhand knot (Figure 7-26) is used to make a knotted rope for a handline, to secure the ends of other knots, and to make stirrups in direct-aid climbing. This knot can also be used to temporarily whip the end of a rope. It is used with other knots.

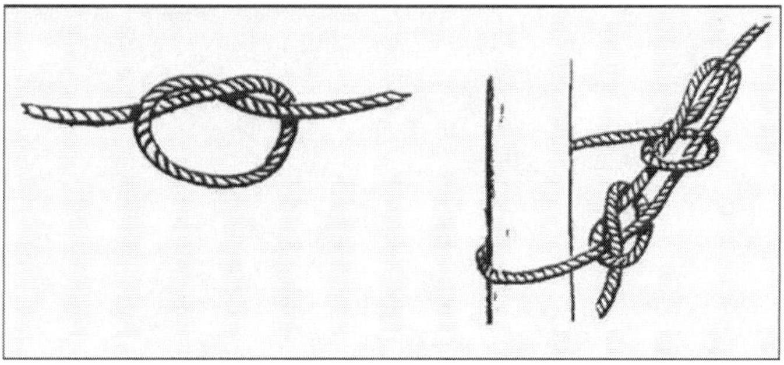

Figure 7-26. Overhand knot.

a. **Tying the Knot.**

STEP 1. Make a loop in the rope.
STEP 2. Pass the working end of the rope through the loop.
STEP 3. Dress down the knot.

b. **Checkpoints.**

(1) Running ends exit the loop in opposite directions.
(2) The knot is dressed down tightly.

7-25. FROST KNOT

The frost knot is used when working with flat webbing (Figure 7-27). It holds better than a standard overhand knot or square knot.

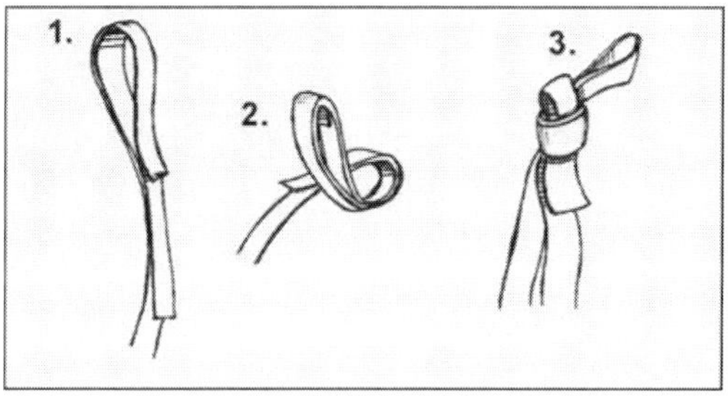

Figure 7-27. Frost knot.

a. **Tying the Knot.**

STEP 1. Lap one end (a bight) of webbing over the other (about 25 to 30 centimeters [10 to 12 inches]).
STEP 2. Tie an overhand knot with the newly formed triple-strand webbing; dress tightly.

b. **Checkpoints.**
(1) The tails of the webbing run in opposite directions.
(2) Three strands of webbing are formed into a tight overhand knot.
(3) A bight and tail exit the top of the overhand knot.

7-26. GIRTH HITCH KNOT

The girth hitch knot is used in tying a runner to a partly driven piton (Figure 7-28).

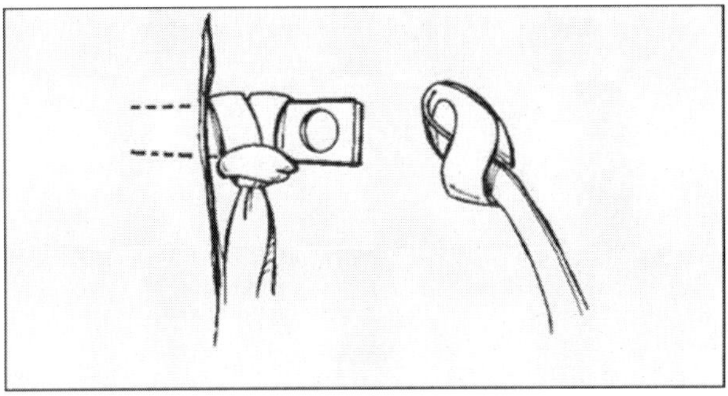

Figure 7-28. Girth hitch knot.

a. **Tying the Knot.**

STEP 1. Bring the standing ends back through the bight of a length of rope or webbing.
STEP 2. Dress the knot tightly.

b. **Checkpoints.**
(1) Two wraps exist with a locking bar running across the wraps.
(2) The knot is dressed tightly.

APPENDIX A

RISK ASSESSMENT

This appendix explains risk assessment for rappel operations. A risk assessement matrix is shown in Table A-1.

A-1. EFFECT

a. **Catastrophic.** Death or permanent total disability, system loss, major property damage.

b. **Critical.** Permanent partial disability, temporary total disability in excess of three months, major system damage, significant property damage.

c. **Moderate.** Minor injury, lost workday accident, compensable injury or illness, minor system damage, minor property damage.

d. **Negligible.** First aid or minor supportive medical treatment, minor system impairment.

A-2. PROBABILITY

a. **Frequent.**

(1) Individual soldier or item—Occurs often in career or equipment service life.

(2) All soldiers exposed or item inventory—Continuously experienced.

b. **Likely.**

(1) Individual soldier or item—Occurs several times in career or equipment service life.

(2) All soldier exposed or item inventory—Occurs frequently.

c. **Occasional.**

(1) Individual soldier or item—Occurs sometime in career or equipment service life.

(2) All soldiers exposed or item inventory—Occurs sporadically, or several times in inventory service life.

d. **Seldom.**

(1) Individual soldier or item—Possible to occur in career or equipment service life.

(2) All soldiers exposed or item inventory—Remote chance of occurrence; expected to occur sometime in inventory service life.

e. **Unlikely.**

(1) Individual soldier or item—Can assume will not occur in career or equipment service life.

(2) All soldiers exposed or item inventory—Possible, but improbable; occurs only very rarely.

A-3. RISK LEVELS

a. **Extremely High.** Loss of ability to accomplish mission.

b. **High.** Significantly degrades mission capabilities in terms of required mission standards.

c. **Medium.** Degrades mission capabilities in terms of required mission standards.

d. **Low.** Little or no impact on mission accomplishment.

RISK ASSESSMENT			HAZARD PROBABILITY				
			Frequent	Likely	Occasional	Seldom	Unlikely
			A	B	C	D	E
EFFECT	Catastrophic	I	***Extremely High***				
	Critical	II			***High***		
	Moderate	III		***Medium***			
	Negligible	IV			***Low***		

Table A-1. Risk assessment matrix.

APPENDIX B

BASIC EQUIPMENT FOR RAPPEL OPERATIONS

The equipment listed in this appendix is the standard basic items required to conduct ground and air rappelling operations.

B-1. GROUND OPERATIONS

For ground operations, the following equipment is needed:

a. Nylon rope, 120 feet (NSN 4020-00-931-8793).
b. Snaplinks (NSN 8465-00-360-0228).
c. Gloves (NSN 8415-00-268-7868).
d. Rope coiling log or standard deployment bag.
e. Rappel seat.
f. Rappel rings (used as expedient anchor points).

B-2. AIR OPERATIONS

For air operations, the following equipment is needed:

a. All items in paragraph B-1.
b. Donut ring (UH-1H operations only) (locally produced TSC item).
c. Safety floating ring.
d. Two 6-foot safety ropes (with snaplinks).
e. Eight 120-foot nylon ropes.
f. Monkey harness for static rappel master.
g. Sixteen snaplinks (UH-60 operations).
h. Twenty-two snaplinks (UH-1H operations).
i. FM communications.
j. Aircraft communications headset.

GLOSSARY

ACRONYMS AND ABBREVIATIONS

AFRM	assistant fast-rope master
AGL	above ground level
AMC	air mission commander
AR	Army regulation
ATP	aircrew training program
attn	attention
BDU	battle dress uniform
CG	commanding general
CGU	modified helicopter safety strap
chemlight	chemical light
cm	centimeter
DA	Department of the Army
D-bag	deployment bag
F	Fahrenheit
FM	frequency modulated; field manual
FRIES	fast-rope insertion/extraction system
FRM	fast-rope master
GSO	ground safety officer
HQ	headquarters
IAW	in accordance with
ICS	interagency communications system
intercomm	intercommunication
LBE	load-bearing equipment
LZ	landing zone
MEDEVAC	medical evacuation
mm	millimeter
mph	miles per hour
MSO	mountaineering safety officer
NCO	noncommissioned officer
NCOIC	noncommissioned officer in charge
NSN	national stock number
NVG	night vision goggles

NWTC	Northern Warfare Training Center
OIC	officer in charge
PIC	pilot in command
POI	program of instruction
POL	petroleum, oils, and lubricants
RATELO	radiotelephone operator
RSO	rappel safety officer
SFC	sergeant first class
SOAR	Special Operations Aviation Regiment
SOP	standing operating procedures
SPIES	special patrol infiltration/exfiltration system
STABO	a system for extracting personnel by helicopter (the combined first letters of the surnames of the five persons who designed the system)
SSO	STABO system operator
TAPO	Technology Applications Program Office
TC	training circular
TM	technical manual
TRADOC	Training Doctrine Command
TSC	training support center
UIAA	Union of International Alpine Association
UH	utility helicopter
US	United States
USAF	United States Air Force
USAIS	United States Army Infantry School
USASOC	United States Army Special Operations Command
VIP	very important person

REFERENCES

DOCUMENTS NEEDED

These documents must be available to the intended users of this publication.

DA Form 2028	Recommended Changes to Publications and Blank Forms. 1 February 1974.
DA Form 5752-R	Rope History and Usage. May 1989.
FM 57-38*	Pathfinder Operations. 9 April 1993.
FM 7-93*	Long Range Surveillance Unit Operations. 3 October 1995.
TC 90-6-1*	Military Mountaineering. 26 April 1989.

READINGS RECOMMENDED

These sources contain relevant supplemental information.

AR 95-3	Aviation: General Provisions, Training, and Standardization, A Resource Management. 27 September 1990.
FM 8-35	Evacuation of the Sick and Wounded. 22 December 1983
FM 57-220	Basic Parachuting Techniques and Training. 1 June 1990.
TM 10-1670-262-12	Operator's and Organizational Maintenance Manual (Including Repair Parts and Special Tools List) for Personnel STABO Extraction System and Cargo Personnel Lowering Anchor Device. 7 April 1988.
USASOC Reg 350-2	Training Airborne Operations. 1 June 1995.
USASOC Reg 350-6	Training Special Operations Forces Infiltration/Exfiltration Techniques. 12 May 1995.
	The Air Assault School Handbook.

*This source was also used to develop this publication.

INDEX

TC 21-24
10 SEPTEMBER 1997

By Order of the Secretary of the Army:

DENNIS J. REIMER
General, United States Army
Chief of Staff

Official:

JOEL B. HUDSON
Administrative Assistant to the
Secretary of the Army
03807

DISTRIBUTION:

Active Army, Army National Guard, and U.S. Army Reserve: To be distributed in accordance with the initial distribution number 115059, requirements for TC 21-24.

FIELD MANUAL
No. 7-85

HEADQUARTERS
DEPARTMENT OF THE ARMY
Washington, DC, 9 June 1987

FM 7-85
RANGER UNIT OPERATIONS

IMPORTANT

U.S. Army Infantry School Statement on
U.S. NATIONAL POLICY CONCERNING ANTIPERSONNEL LAND MINES

TABLE OF CONTENTS

 DISTRIBUTION RESTRICTION: Approved for public release; distribution is unlimited.

PREFACE

This publication contains doctrinal guidelines for the employment and training of ranger units. The intended audience is intermediate and higher level commanders, staffs of those headquarters, and newly assigned officers and NCOs of the ranger regiment. It is to be used with all other published US doctrine concerning NBC operations and the climatic or geographic influences on combat. The fundamentals apply to the employment of ranger units on worldwide operations and across the spectrum of conflict from low to high intensity warfare. This includes all aspects of the integrated battlefield: nuclear, biological, and chemical. It is not intended to apply to the employment of long-range reconnaissance patrol (LRRP) companies, even though such units carry the designation "ranger." These LRRP companies, as well as the long-range surveillance company (LRSC) (corps level) and the long-range surveillance detachment (LRSD) (division level), are organized, trained, and equipped for long-range reconnaissance in enemy territory.

The scope of this publication addresses the mission, organization, equipment, capabilities, limitations, planning, training, operations, and logistical support of ranger units. It describes how the ranger regiment and its subordinate battalions train, receive taskings, plan missions, task-organize, deploy, and integrate supporting elements to perform the mission. Once in combat, the ranger regiment maneuvers to accomplish the mission in the manner of other light infantry units. Ranger units make full use of published light infantry doctrinal guidance. This manual will repeat procedures or techniques contained in other publications only if necessary for understanding or continuity.

This publication does not contain everything the ranger needs to know to successfully execute a unit mission. Rather, it supplements unit training and the ranger unit commander's guidance; it does not supersede it.

The proponent of this publication is HQ, TRADOC. Submit changes for improving this publication on DA Form 2028 (Recommended Changes to Publications and Blank Forms) and forward it to Commandant, US Army Infantry School, ATTN: ATSH-B-ID, Fort Benning, GA 31905.

Unless otherwise stated, whenever the masculine gender is used, both men and women are included.

CHAPTER 1

Introduction

The ranger regiment is a major component of the US Army's special operations forces. It is a unique light infantry unit tasked to conduct special military operations in support of national policies and objectives. These operations require highly trained, well-disciplined units capable of employment in any environment, either alone or in concert with other military forces. Within this publication, the term *ranger force* describes any size force consisting mainly of members of the ranger regiment and led by a member of the ranger regiment's chain of command. A ranger force may be a TOE unit or it may be a specially organized task force for a specific mission.

1-1. Mission.

a. The mission of the ranger regiment is to plan and conduct special military operations. These operations are conducted by specially trained, equipped, and organized forces against strategic or tactical targets in pursuit of national military, political, economic, or psychological objectives. They may support conventional military operations or they may be performed independently when conventional forces cannot be used. (See Figure 1-1.)

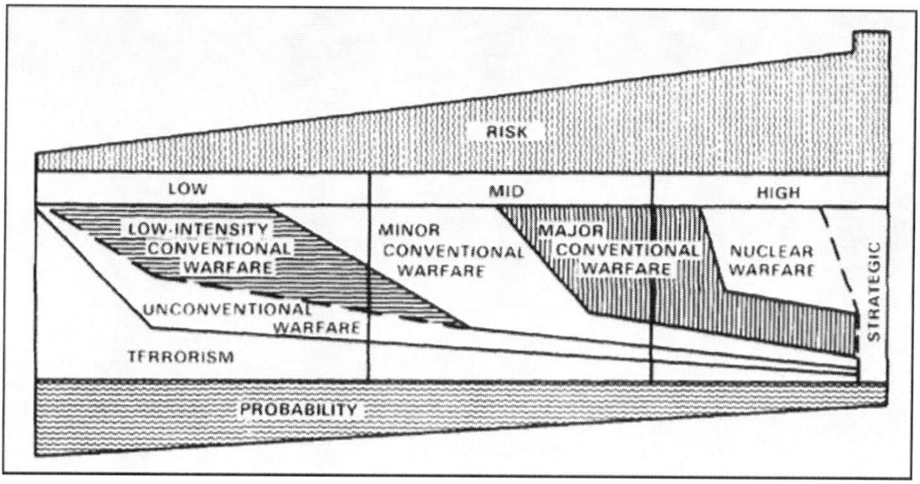

Figure 1-1. Spectrum of conflict.

b. Special military operations conducted by the ranger regiment include strike operations, usually deep penetration, and special light infantry operations. Strike operations include raids, interdiction, and recovery operations. Special light infantry operations include many of the light infantry missions assigned to airborne, air assault, or light infantry battalions and brigades. These operations are conducted in support of the AirLand Battle at all levels of intensity.

1-2. Purpose and function.

The ranger regiment provides the national command authority (NCA) the ability to move a credible military force quickly to any region in the world. The regiment uses the entire spectrum of intelligence support, from national systems to organic assets. Ranger units maintain a readiness posture that supports their immediate commitment to battle once deployed. They are often tailored for specific missions and may require augmentation from external sources. Tactical mobility may be augmented by USAF or Army special operations aviation (SOA) aircraft.

a. The ranger regiment conducts light infantry operations on the integrated battlefield as well as in low intensity conflicts. Special light infantry operations are conducted to accomplish certain contingency missions during peacetime or as part of the echelons above corps (EAC) battle.

b. The regiment conducts both strategic and operational missions. Its efforts are combined into an overall plan to destroy, delay, and disorganize the enemy, or to cause him to divert his attention and combat forces to rear area security. The limited number of ranger units and the diverse targets dictate a careful assignment of missions.

c. The mission, enemy, terrain, troops and time available (METT-T) factors govern the command and control arrangement under which the regiment operates. The ranger regiment or separate ranger battalions are normally assigned to the headquarters whose area of responsibility includes the regiment's operational area(s). The strategic or operational value of potential targets means that ranger units are normally employed at no lower than corps level. Employment at EAC is routine. The EAC headquarters employing ranger units could be the commander in chief's (CINC's) special operations command or field army. The controlling headquarters could also be that of a joint task force (JTF) or the Army component command (ARFOR) of a JTF. The ranger regiment's organization, communications equipment density, and training programs do not support operational level reconnaissance missions.

d. Ranger battalions are not oriented to a specific theater. Current force structure, contingency plans, and training needs (see Appendix G) preclude committing battalions to one region.

1-3. Fundamentals of ranger operations.

a. The success of an operation by a ranger unit depends on the observance of the fundamentals of the US Army's AirLand Battle doctrine (see FM 100-5). The ranger regiment fights the enemy as a light infantry force. It follows infantry doctrine and observes all the basic rules of infantry operations.

b. Along with initiative, depth, agility, and synchronization, ranger operations require--

(1) Detailed planning and coordination that allow the ranger unit to discern and exploit the enemy's weaknesses while avoiding his strength. Both permit the evaluation and use of information gathered from all sources. Detailed coordination integrates all supporting units and services. During planning, the ranger unit conducts pre-mission training, briefings, and rehearsals for all personnel. The use of special mission equipment is practiced and perfected.

(2) Decentralized execution IAW mission orders and the commander's intent. Special operations forces must use individual and unit initiative. Mission-type orders give the ranger force commander the flexibility to take advantage of opportunities on the battlefield.

(3) Surprise, achieved through the ranger unit's ability to move by uncommon means, along unexpected routes, and over rough terrain. Ranger units normally conduct operations during poor weather and reduced visibility, aided by night vision devices. This adds to the attainment of surprise.

(4) Survivability, achieved by using the classic infantry combat techniques of stealth and concealment. The ranger unit engages the enemy at the time and place of its own choosing. It takes full advantage of terrain, and destroys or suppresses enemy weapons. The ranger unit seeks to destroy or neutralize the enemy's command and control systems, his surveillance assets, and his mobility assets. Survivability is enhanced by rapid mission accomplishment and a prompt departure from the objective area.

(5) Mobility, speed, and violence of execution that allow the ranger unit it to close quickly on the objective area and complete the mission before the enemy can react. The speed at which events take place confuses and deceives the enemy as to the intent of the ranger unit. This forces the enemy to react rather than to take the initiative. Tactical mobility lets the ranger unit break contact and withdraw from the objective area.

(6) Shock effect, which is a psychological advantage achieved by the combining of speed and violence with the precision of the ranger attack. The ranger unit strives to apply its full combat power at the decisive time and place, and at the point of the greatest enemy weakness. The ranger unit strives to achieve maximum physical and psychological effect on the enemy by exhibiting aggressiveness and reasoned audacity.

(7) Multiple methods of insertion and attack, trying not to repeat operations thus decreasing the chance the enemy will detect a pattern. This is achieved through imaginative training and planning.

(8) Deception, practiced by the ranger unit during all phases of its operations, from deployment through the insertion phase, to the actions in the objective area and extraction. The enemy is kept unaware of the ranger unit's intentions, is fed conflicting and wrong information, and is kept from knowing the size or mission of the ranger force. The ranger force commander makes full use of ruses, feints, false insertions, electronic countermeasures, and dummy transmissions. Concurrent military operations by US or allied forces may be used as cover for ranger operations. Ranger units are careful to hide every possible aspect of an operation and to disguise those that cannot be hidden. While in the operational area, or when departing any location, ranger units ensure that no material is left that could provide the enemy with information concerning the ranger force.

(9) Audacity, achieved by a willingness to accept risk. The ranger force commander considers what the enemy expects the unit to do and the actual ranger abilities, He then chooses a course of action that may confuse the enemy, while remaining within the capability of the ranger unit. This fundamental is combined with deception and surprise to disrupt the enemy command and control.

c. Detailed planning and coordination finds the enemy. Surprise, mobility and speed, variety, deception, and audacity combine to shock and disorient the enemy, fixing him in place. The violence and precision of the ranger attack finishes the enemy while ensuring the ranger force survives.

1-4. Capabilities and limitations.

a. The ranger regiment has the following capabilities:

(1) Deploying quickly to conduct operations on all types of terrain and in all kinds of weather.

(2) Establishing a credible American presence in any part of the world to show US interest or resolve.

(3) Infiltrating and exfiltrating an area of operations, and assaulting an objective by land, sea, or air.

(4) Conducting strike operations to include raids, personnel and equipment recovery operations, and interdiction of key areas.

(5) Conducting special light infantry operations to include seizing and securing airfields, communications centers, command and control facilities, and key bridges; and other special light infantry operations.

(6) Performing short-duration reconnaissance of assigned ranger objectives for the ranger force commander.

(7) Operating for up to three days without resupply, and for longer periods when provided with accompanying or airdropped supplies.

(8) Providing liaison, communication, and coordination personnel and equipment to integrate the deployed ranger force into the logistical, intelligence, and operational system of the theater or joint task force (JTF) commander.

(9) Assuming operational control, for a limited time, of other US military forces such as engineers or infantry, airborne, or air assault battalions.

(10) Providing the focal point for all-source intelligence support to attached and assigned units of the regimental task force.

(11) Conducting limited combat operations under conditions of chemical, nuclear, or biological contamination.

b. The ranger regiment has the following limitations:

(1) Limited capability against armored or motorized units in open terrain.

(2) No organic transportation.

(3) Limited sustained combat capability due to the shortage of organic combat support and combat service support elements.

(4) Limited organic air defense weapons.

(5) Limited organic indirect fire support.

(6) No casualty evacuation capability.

(7) Reconstitution and retraining needed to replace combat losses.

1-5. Employment considerations.

a. Ranger units are characterized by the quality, motivation, training, and individual skill of their members. This produces units with superb collective abilities, able to adapt well to changing, complex situations.

b. Ranger units can conduct either deliberate or quick-response operations.

(1) Deliberate operations rely on careful planning, reconnaissance and surveillance of the target area, deception, secrecy, thorough preparation and rehearsals, and violent execution. A deliberate operation aims to complete the mission even though the enemy may have heavy forces on or near the objective area. Deliberate operations allow for detailed planning, evaluation, rehearsal, and coordination before insertion. A deliberate operation is likely to succeed against targets that the enemy has protected in depth, that have strong natural defenses, or that need a detailed and long insertion process.

(2) Quick-response operations rely on the high level of training and readiness of the ranger regiment to execute a mission before the enemy can react. These operations are conducted when there is little time for long, detailed planning. They rely on set procedures set forth by the ranger regiment and its supporting elements. A quick-response operation aims to complete the mission before an enemy can react. This type of operation may be chosen due to the time-sensitive nature of the target, political or military goals, the time frame of other operations, or the increased chance of enemy detection.

(3) Whether an operation is to be deliberate or quick response is often a difficult and time-sensitive decision by a high-level command authority. Decision-makers must consider the enemy's strength in the area, his intentions, his ability either to reinforce or to alter the target area, and the consequences of success or failure of diplomatic or military initiatives in related areas. The ranger unit commander tries to use the existing time, manpower, and resources to complete a detailed and coordinated plan. He refines that plan up to the insertion into the objective area.

c. Ranger units train to operate in any environment or weather condition. They regularly perform operations during periods of limited visibility. Ranger units maintain a high state of physical fitness and often train in close quarters combatives. In addition to completing advanced marksmanship training with standard US weapons, each member of a ranger unit trains with many foreign weapons. Ranger units are trained to operate on urbanized terrain, becoming specialists in entry and clearing techniques and quick-fire methods, especially during periods of limited visibility.

d. Because ranger units have limited vehicles, logistics operations capability, indirect fire support, and heavy weapons systems, they are not designed for continuous operations. During all phases of operations and training, ranger units need responsive external support.

e. Ranger units are normally employed against targets and under conditions that need their unique skills. Although targeting priority is set by the overall commander, ranger units are not normally assigned missions that can be done by conventional aerial bombardment or by other units.

f. Ranger units are oriented toward offensive operations. They are not normally employed as a rear area protection force. Although the ranger regiment has a small reconnaissance unit, ranger units do not normally conduct long-range reconnaissance missions. The structure, communications, and training of the ranger unit do not prepare it for LRRP missions. Ranger units engaged in strike and special light infantry operations have a secondary mission to collect and report combat information.

g. Ranger units can be deployed worldwide when US military presence or participation with a host national military activity would serve US interests. This deployment shows a readiness to commit forces into a threatened area or proves US national resolve. After the deployment, other activities include staging operations, rehearsing combat operations, securing base areas for use and deployment of other forces, and so on that provide a clear signal of US intent. Ranger units are not trained or organized to provide mobile training teams (MTT) to train indigenous forces. The US Special Forces or other special operation forces are trained to conduct such security assistance operations.

h. Ranger units can serve as an example to a host country and provide limited military advice and training. They will normally be augmented with linguists and technicians to increase their abilities. The regiment would normally still function as a unit. Most of its assistance would be through short-term, high-impact, unit-oriented operations. They would not be long-term individual efforts associated with advisory-type activities.

i. Ranger units may be deployed to engage in combined training exercises with allies. This enhances US national image by demonstrating the outstanding abilities of the American ranger. These activities may include ranger, light infantry, airborne, air assault, or amphibious operations.

j. Ranger units use standard US nuclear, biological, chemical (NBC) warning; detection; protection; and decontamination equipment and doctrine when operating on the integrated battlefield.

1-6. Command and control.

a. **External.**

(1) The ranger regiment is a key component of the US Army's special operations forces. The other elements of special operations forces are special forces (SF), psychological operations (PSYOP), civil affairs (CA), and special operations aviation.

(2) The 1st Special Operations Command (1st SOCOM) is responsible for command of all active component special operations forces elements, to include the ranger regiment, in peacetime. The 1st SOCOM has command and control of all assigned special operations forces, less operational command (OPCOM) of special operations forces units forward deployed. (See Figure 1-2.) The ranger regiment, while based in the continental United States (CONUS), is under the control of 1st SOCOM. During peacetime, the OPCOM of deployed ranger units is through channels chosen by the NCA through the Joint Chiefs of Staff (JCS) to the unified command.

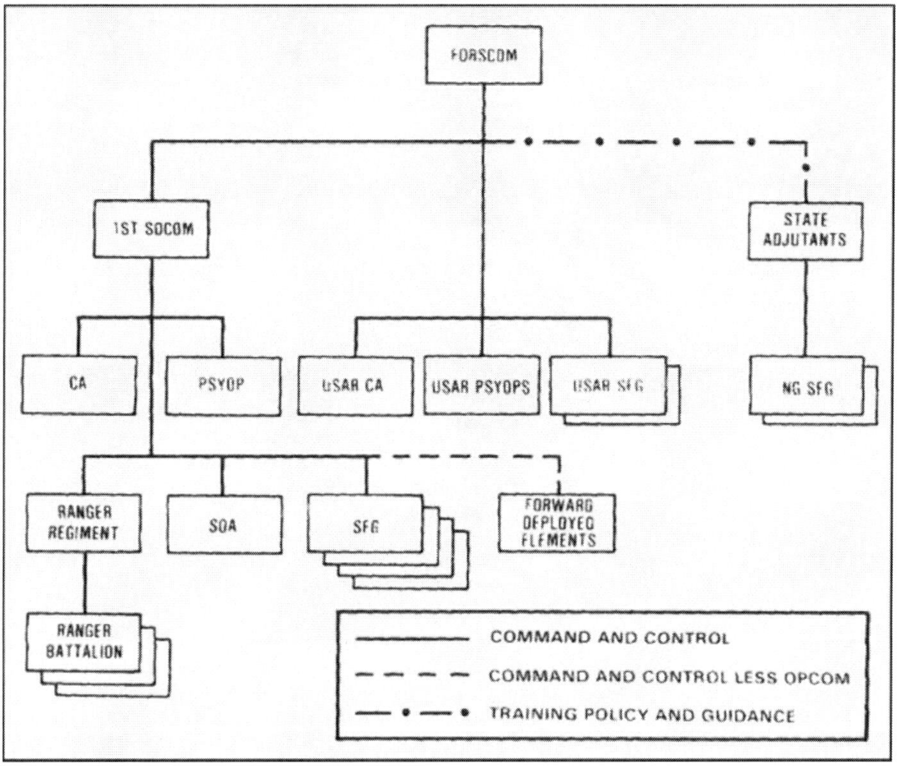

Figure 1-2. Command and control of Army Special operations forces during peacetime.

(3) In support of national policy, the NCA may direct contingency operations involving US

forces. These are politically sensitive military operations normally characterized by the short-term rapid projection or employment of forces in conditions short of conventional war. One aim of the NCA is to conclude them without going to war. The flexibility and effectiveness of the ranger regiment make it possible to use military force quickly to complete the mission and limit the spread of conflict. Command and control of special operations forces in a contingency operation is usually by a special operations task force (SOTF). (See Figure 1-3.) The SOTF and designated special operations forces are under the OPCOM of the highest level of command responsible for the operation. This command may be the unified command responsible for the contingency area (Option A), or a JTF designated by the NCA through the JCS (Option B). If no SOTF is formed, the ranger regiment could be OPCOM to the Army component commander (Option C). The ranger regiment or its battalions are not normally placed OPCOM to a division. A division's area of influence rarely contains appropriate targets. The division also lacks the assets to adequately support ranger operations. Employment of a ranger unit by a corps-level command is normally on a case-by-case basis. The theater or JTF commander retains centralized control over the mission assigned to the ranger unit.

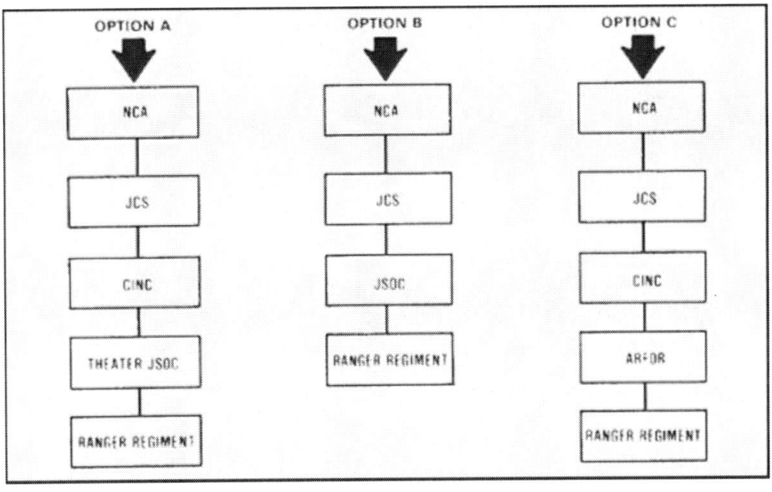

Figure 1-3. Command and control of Army Special operations forces during contingency operations.

(4) The 1st SOCOM is responsible for the deployment of the ranger force to the theater area. The headquarters having OPCOM provides the all-source intelligence, secure communications, insertion and extraction assets, and logistical support needed. The responsible commander in chief, through the Army component commander, provides support during the contingency operation.

(5) During wartime, ranger units are deployed as strategic assets to the theater of operations.

Strategic mission support is provided to the unified commander (CINC) from the NCA through the Joint Chiefs of Staff. It may involve the execution of sensitive operations. The employment of ranger units is controlled at the highest level of command directly responsible for the mission. Ranger units are assigned to a REMAB based on mission, force protection, and OPSEC considerations. The employing command and other elements provide timely and accurate intelligence, communications, logistical, and administrative support. Command and control of these units conforms to the provisions of JCS Pub 2. (See Figure 1-4.)

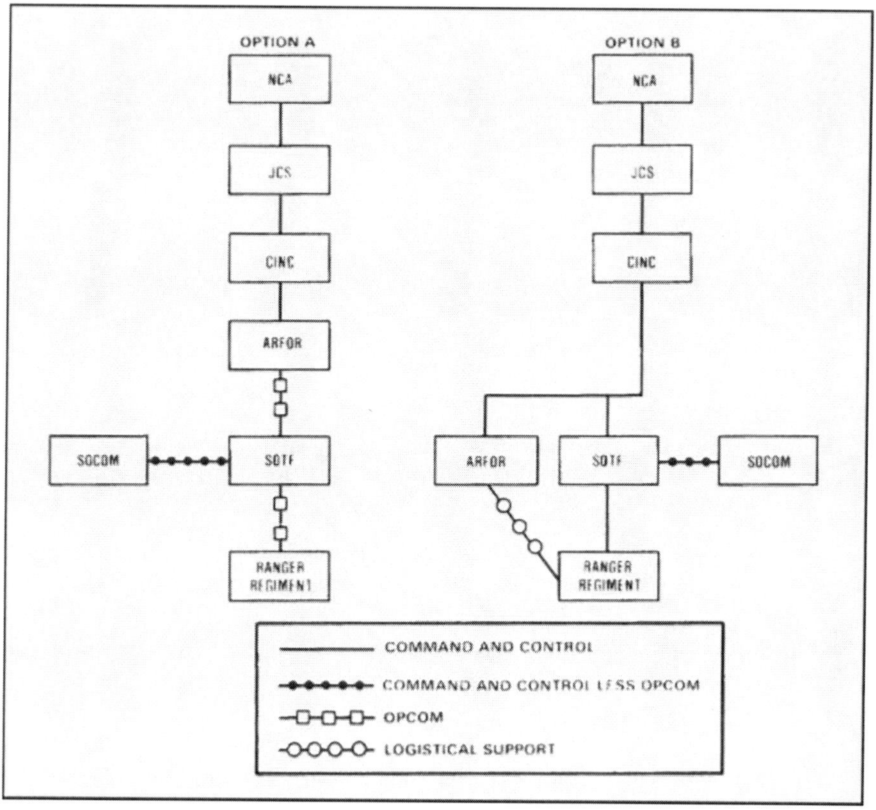

Figure 1-4. Command and control of Army Special operations forces during wartime.

(a) A theater joint special operations command (JSOC) is subordinate to the unified command to supervise the command and control, employment, and support of all US and allied special operations forces. The JSOC is a key element in the command and control of ranger units performing strike missions or special light infantry operations (Option A).

(b) The JSOC may be placed on a lateral line with other in-theater service components (Option B). This can occur when ranger units are -employed far from conventional forces and against strategic targets. It can also occur when the JCS or a CINC forms a JTF to conduct a limited-duration operation in an area where ranger units and conventional forces may operate near each other, or where operations must be coordinated on a regional level.

(c) Under the Army special operations command (ARSOC) concept, 1st SOCOM deploys as a command and control headquarters within the theater. When this occurs, the command relationships will be as shown in Figure 1-5.

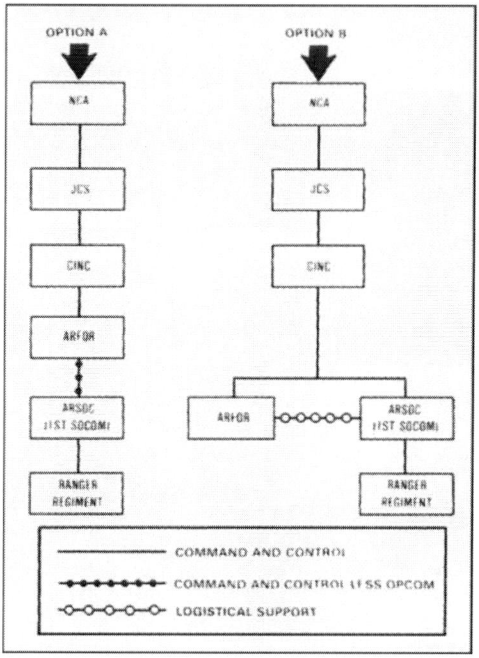

Figure 1-5. Command and control of Army Special operations forces under the ARSOC concept.

(d) Recently passed legislation will modify existing high level SOF command and control arrangements. All US Army, USN, and USAF special operations forces will be consolidated into joint commands at both national and theater level. The ranger regiment will continue to operate under the command of the 1st SOCOM, while increasing its ability to conduct joint operations.

(6) Ranger units conducting strike or special light infantry missions are not normally offered for transfer of authority (TOA) to allied forces. However, this will not preclude allied commands and US elements from combined planning and targeting for ranger missions.

(7) Requests for ranger units, from either US or allied commands, go through normal command channels to the CINC. The CINC, through the JSOC, is responsible for the employment of all special operations forces, including ranger units. This provides flexible response to JCS-directed missions, ensures quick response to the CINC's needs, enhances the joint employment of special operations forces, and lessens the chance of overcommitting and degrading the ranger force.

(8) Command and control of the ranger regiment and its battalions is IAW AirLand Battle doctrine. Before commitment to a mission, command and control is kept at a level where the unit's unique skills can be applied worldwide. Once committed, the ranger unit is placed under the operational control of the command that is responsible for the mission. This normally corresponds to the command whose area of influence includes the target area. Ranger task forces may operate directly under the authority of the senior American representative in the country. Command and control arrangements may vary according to the theater involved and METT-T considerations.

(9) The command and control of special operations is centralized and begins at the highest level of the military and civilian decision-making process. The specific command relationships are normally based on the mission.

b. **Internal.**

(1) The ranger regimental headquarters is organized to operate like that of a brigade headquarters. In addition to commanding and controlling all three ranger battalions, the regimental headquarters can assume operational control (OPCON) of conventional combat and combat support units, and other special operations forces for limited periods. It provides the regimental commander with a well-balanced responsive staff. It is structured as an operational headquarters that can deploy to the mission area and act as the ground tactical headquarters.

(a) The regimental headquarters prepares for combat by forming two small command and control groups. Each of these can deploy to an objective area and control combat operations. Other elements of the staff operate the tactical operations center (TOC) and the logistical operations center (LOC). These may be collocated depending on the situation. The TOC and LOC are normally located well behind the FEBA, at the REMAB or at the intermediate staging base (ISB). The ranger regiment also has two small liaison teams consisting of LNOS, staff representatives, and communications elements. (See Appendix B.)

(b) When a single ranger battalion is committed to an operation, the regimental commander normally exercises control from the TOC. A liaison and communications team is detached from the regimental headquarters to augment the ranger battalion. This team is attached to the deployed battalion until the end of the mission.

(c) If two or more ranger battalions are used in a single operation, the regimental commander normally deploys both command and control groups. He commands from the objective area, The use of two small, mobile control elements makes up for the possible loss of the regimental commander and the primary control element.

(d) Whenever the regimental commander is controlling an operation from the REMAB or the objective area, he normally provides a liaison team with a communications element to the higher headquarters exercising operational control. This cell can provide ranger representation to the controlling staff, operations interface, logistics interface, fire support planning and coordination, intelligence analysis and dissemination, and communications between the higher headquarters and the ranger battalion.

(2) The ranger regimental headquarters can serve as a ranger or special operations force task force headquarters. With augmentation, it can function as the Army special operations force command element. The regimental headquarters can also function as the Army component command of a JTF.

1-7. Communications.

a. Ranger missions need unique and dedicated communication support assets. Along with the ranger regiment's organic communications means, communications support assets are organized within the 1st SOCOM that-can deploy in support of the ranger regiment. This ensures that there are reliable communications to all elements of the SOTF. A special operations communications battalion (SOCB), deployed by the 1st SOCOM, provides more communications systems to support the special operations plan. The SOCB does not replace or duplicate existing in-theater communications. It augments the CINC's communications with the US Army's portion of the critical special operations force command, control, and communications link unique to the SOTF.

b. The ranger regiment's communication platoon can provide secure, long-range communication links from the CINC to a deployed ranger force. It can also enter the SOCB and the joint communications support element (JCSE) nets to provide a direct link with the NCA. Its equipment consists of four light vehicles equipped with various multiple communications systems. The two regimental liaison teams use one of these vehicles each. The TOC and LOC each use one.

c. Secure satellite communications (SATCOM) and radios with AM or FM capability are the main means of communication within the ranger regiment. These are also used for communications within the ranger battalions. The regimental commander has two light vehicles equipped with multiple communications systems. One of these is used to support each of the two mobile command and control elements of the regimental headquarters. Some of the radio systems can be dismounted and man-packed.

d. The ranger regimental headquarters has both man-packed and vehicular tactical satellite communication systems. Each of the ranger battalions has one base station and three man-packed systems.

(1) The man-packed radio is battery-operated. It provides two-way communications in both line-of-sight (mobile) and satellite (at-halt) modes. The digital message device group (DMDG) is the input and output device for data transmission. Secure voice communications is achieved using various encryption devices.

(2) The vehicular net control station uses the basic receiver/ transmitter unit installed on a vehicle. Each can serve up to 15 terminals in a communications net.

e. The reconnaissance platoon is equipped with multiple, man-packed communications systems. These let its teams operate in most of the communications nets within the regiment. Each five-man reconnaissance team has this communications ability, as does the platoon headquarters.

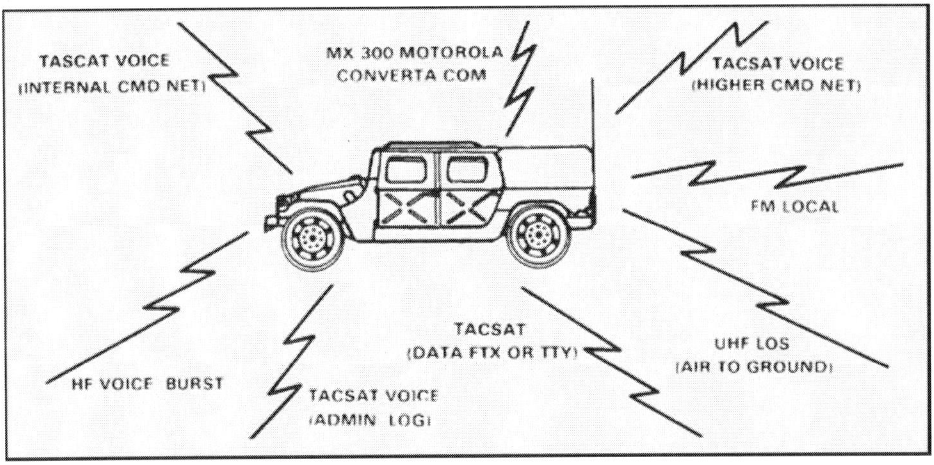

Figure 1-6. Communications capabilities within the regimental headquarters.

CHAPTER 2

Organization and Equipment

The ranger regiment is a specially organized and trained light infantry force consisting of about 2,000 personnel. It includes a regimental headquarters and headquarters company and three ranger battalions. (See Appendix A.) The regiment can operate as a single unit, separate battalions, or separate companies on specific missions. Independent missions are not normally assigned to ranger platoons.

2-1. Headquarters and headquarters company.

The HHC of the ranger regiment consists of 130 personnel, the majority being senior noncommissioned officers, commissioned officers, or soldiers with special skills to support the operations of the ranger battalions. (See Appendix A.) The HHC consists of two elements: the regimental headquarters and the headquarters company (minus).

2-2. Regimental headquarters.

a. The regimental headquarters consists of the regimental commander and his staff. The primary staff is composed of five sections: personnel (S1), intelligence (S2), operations and training (S3), supply (S4), and civil affairs (S5). These staff sections perform the standard functions and are supervised by the executive officer (XO). The regimental commander has a deputy commander and a special staff consisting of a communications officer, fire support officer, surgeon, and staff judge advocate. The S2 officer has a large intelligence analysis and counterintelligence section. The S3 section has extra liaison officers and assistant operations officers to coordinate the widespread and complex ranger missions. A United States Air Force (USAF) staff weather officer and a tactical air control party (TACP) are permanently attached.

b. The regimental headquarters:

- Analyzes assigned missions.
- Plans, coordinates, and controls training and operations.
- Coordinates the infiltration, resupply, exfiltration, and debriefing of the reconnaissance teams.
- Provides liaison support.
- Provides administrative, medical, NBC, and logistical planning and coordination support to the committed ranger battalions.
- Coordinates special support requirements (rigger, transportation, communications security [COMSEC]) for the ranger battalions.

2-3. Regimental headquarters company.

The regimental headquarters company (minus) consists of the company commander and his administrative personnel, plus the limited combat support (CS) and combat service support (CSS) elements assigned to the regiment. These are the fire support element, the communications platoon, the reconnaissance platoon, and the medical treatment team.

a. The company headquarters:

(1) Provides command and control of assigned elements.

(2) Provides administrative, personnel, logistical, and maintenance support to the assigned elements.

b. The fire support element:

(1) Plans fires to support the commander's ground tactical plan.

(2) Coordinates the integration of all available fire support means (mortars, artillery, armed helicopters, close air support, naval gunfire) to support mission execution.

c. The communications platoon (see Appendix A):

(1) Provides secure SATCOM teletype (TTY) and facsimile transmissions (FAX) abilities at the remote marshalling base or intermediate staging base (REMAB or ISB).

(2) Provides secure SATCOM, TTY, and FAX to the regimental headquarters and ranger battalions in the objective area.

(3) Provides cryptographic support to the regimental headquarters.

(4) Provides limited repair of radio and cryptographic equipment.

d. The reconnaissance platoon (see Appendix A):

(1) Trains, equips, and controls the employment of three reconnaissance teams.

(2) Infiltrates the objective area by parachute (high-altitude, low-opening [HALO]; high-altitude, high-opening [HAHO]; static line), helicopter, fixed-wing aircraft, self-contained underwater breathing apparatus (SCUBA), small boat, or other means.

(3) Remains undetected in the objective area for up to five days.

(4) Performs reconnaissance.

(5) Performs demolition target analysis.

(6) Uses small boats and inflatable rafts.

(7) Emplaces and monitors unattended ground sensors.

(8) Uses night image enhancement devices.

(9) Uses photographic equipment.

(10) Emplaces electronic target acquisition and designation devices.

(11) Collects specific combat information to satisfy priority intelligence requirements (PIR).

(12) Reports information using secure, long-range communication equipment.

Figure 2-1. Reconnaissance platoon communications.

(13) Performs drop zone (DZ) and landing zone (LZ) selection, marking, and reception duties.

(14) Escapes and evades enemy areas.

(15) Performs limited attacks or ambushes when tasked by the regimental commander.

(16) Reports weather in the objective area.

e. The medical treatment team:

(1) Provides medical support to the regimental headquarters.

(2) Coordinates required medical support and evacuation.

2-4. Ranger battalion.

a. The ranger battalion is organized and equipped to conduct both special and conventional combat operations. It contains organic combat, CS, and limited CSS elements. It consists of a battalion headquarters, a headquarters company, and three ranger rifle companies. (See Appendix A.)

b. The ranger battalion is the combat element of the ranger regiment. It can conduct independent combat operations or subdivide into its separate ranger rifle companies and their subordinate elements who can conduct semi-independent operations.

c. The ranger battalion headquarters and headquarters company (see Appendix A) includes:

(1) **Battalion headquarters.** The battalion headquarters consists of the commander; the executive officer; the command sergeant major; the S1, S2, S3, S4, and S5 officers; the chaplain; the communications officer; the fire support officer; and two radio operators.

(2) **Battalion headquarters section.** The battalion headquarters section consists of the S3 air and assistant S8, an assistant S2, an intelligence sergeant and a senior intelligence analyst, two operations sergeants and an operations assistant, a personnel staff noncommissioned officer (PSNCO), the chaplain's assistant, two clerk typists, two radio operators, a chemical officer and NCO, a personnel administration center (PAC) NCO, three PAC clerks, and a legal specialist. There is a USAF tactical air control party permanently attached.

(3) **Communications section.** The communications section consists of one telecommunications technician, one tactical communications chief, two combat signal radio team chiefs, one tactical communications systems supervisor, three radio operators, one switchboard operator, one single channel radio operator, and two equipment mechanics. The section provides communications support and associated maintenance.

(4) **Battalion medical section.** The battalion medical section consists of one general medical officer, one physician's assistant, two medical specialists, and twelve aidmen. The section provides unit-level medical support and trains selected members of the battalion in advanced medical procedures.

(5) **Battalion support platoon.** The battalion support platoon consists of a battalion support platoon leader, an ammunition NCO, and a food service section. The platoon provides limited CSS for assigned and attached units. This platoon does not have organic transportation. The battalion food service section consists of a food service sergeant (E7), a first cook (E6), and eight cooks.

(6) **Fire support team headquarters.** The fire support team (FIST) headquarters is supervised by the battalion fire support officer. It consists of the fire support sergeant and two fire support specialists. The headquarters controls three 9-man fire support teams. These teams are normally with the ranger rifle companies in a habitual support relationship. When operations are conducted near water, the fire support team headquarters may be augmented by spotter teams from the naval shore fire control party.

(7) **Company headquarters.** The HHC company headquarters consists of a commanding officer, an executive officer, a first sergeant, a supply sergeant, and an armorer.

d. The ranger rifle company (see Appendix A) includes a company headquarters, three rifle platoons, and a weapons platoon. Each company can conduct semi-independent operations for a short time when augmented by assets from the ranger battalion headquarters company.

(1) **Ranger rifle company headquarters.** The ranger rifle company headquarters consists of a commanding officer, an executive officer, a first sergeant, a supply sergeant, a tactical communications chief, two radiotelephone operators, an armorer, and an NBC NCO. It also has a two-man sniper team equipped with a sniper weapon. It normally has one of the 9-man FISTs attached from the ranger battalion headquarters company. Elements of this FIST may be further attached to the ranger rifle platoons.

(2) **Ranger rifle platoon.** The ranger rifle platoon consists of a headquarters section, three rifle squads, and a machine gun squad. The headquarters section personnel include a platoon leader, a platoon sergeant, and a radiotelephone operator. Each rifle squad consists of a squad leader and two balanced fire teams of four men each. The machine gun squad consists of a squad leader and three M60 machine gun crews of three men each. The machine gun squad is normally employed as a single element to provide quick, accurate, long-range supporting fire. One or more crews may be attached to other squads within the platoon.

(3) **Ranger weapons platoon.** The ranger weapons platoon consists of a headquarters section, a mortar section, and an antiarmor section. The headquarters section consists of a platoon leader, a platoon sergeant, and a radiotelephone operator. The mortar section has two 3-man 60-mm mortar squads. The mortar squads are normally employed as a section. They may be attached by section or squad to other platoons. The antitank section normally consists of a section sergeant and three 3-man 90-mm recoilless rifle or Dragon missile teams. They are normally employed as a section. The antiarmor section may also be attached by section or team to other platoons.

2-5. Special teams and elements.

The ranger regiment forms special teams to do critical tasks. These teams are made up of members of the unit and are used when needed. Examples of these special teams and elements are--

a. SCUBA teams whose members are graduates of the Special Forces Underwater Operations School. They conduct special insertion and reconnaissance duties.

b. HALO teams whose members are graduates of the Special Forces Freefall School and who conduct high-altitude, low-opening parachute operations.

c. HAHO teams whose members are also graduates of the Special Forces Freefall School and who conduct high-altitude, high-opening parachute operations, allowing offset parachute insertions.

d. Scout swimmer teams whose members are specially trained by the Landing Force Training Command (United States Marine Corps [USMC]) to conduct operations along coastlines, waterways, and inland waters. These teams conduct special insertion, reconnaissance, and security tasks often in support of amphibious operations.

e. Sniper teams, whose members are authorized by the table of organization and equipment (TOE), are used for both conventional and special operations. Along with the two-man sniper teams in the rifle company headquarters, each rifle platoon has two trained snipers.

f Demolition teams whose members are trained in advanced demolition and incendiary techniques and the use of special explosives,

g. Air defense teams trained in the use of man-portable air defense system (MANPADS) weapons. Such teams give the ranger battalion limited self-defense against air attack.

h. Liaison teams that are used by the ranger regimental commander to augment the existing command and control links from the employing unit headquarters to the ranger regiment. These teams are placed where they are most needed and may assist the higher headquarters staff in planning and target selection, as well as performing other liaison duties. See Appendix B for a detailed listing of the personnel and equipment included in the ranger liaison teams.

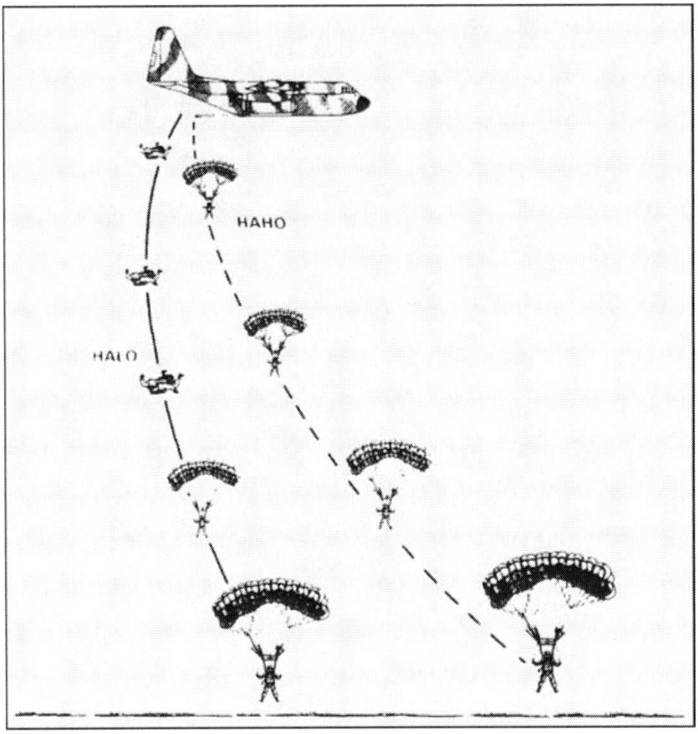

Figure 2-2. HALO and HALO parachute operations.

i. The ranger support element (RSE).

(1) The RSE is a special support element provided by host installation commanders at CONUS posts. It consists of elements and teams from either TOE units stationed at the installation or from the table of distribution and allowances (TDA) organizations under the command of the installation commander.

(2) The RSE provides the necessary support to ensure a timely and efficient deployment of the ranger battalion. During CONUS deployments, the RSE is responsible for both the departure and the reception support provided the battalion. During OCONUS deployments, the RSE is responsible for initial deployment and continuing support until the deploying ranger force is OPCON to the supported unified command or JTF. The responsible theater commander assumes support of the ranger battalion at the REMAB or the ISB if outside continental United States (OCONUS). On deployments that require the ranger battalion to move directly to the objective area from CONUS, the RSE continues support of the ranger battalion until released, and is prepared to accept considerable augmentation assets. The RSE provides the following support:

(a) Furnishing meals that are either served by a supporting unit in an established facility or taken to the ranger battalion at an isolated location.

(b) Issuing supplies from existing stocks, or from contingency items stored for use by the ranger battalion.

(c) Transporting supplies, food, personnel, and equipment.

(d) Palletizing equipment and ammunition for air movement and airdrop.

(e) Rigging vehicles and equipment for airdrop.

(f) Refueling vehicles and equipment.

(g) Issuing barrier and construction material for building of rehearsal sites.

(h) Helping in the construction of rehearsal sites and targets.

(i) Securing, storing, moving, and issuing ammunition and other Class V items.

(j) Maintaining ranger weapons and equipment, to include communications equipment. This is normally done by maintenance contact teams.

(k) Setting up and operating the RSE emergency operations center.

(l) Securing the REMAB if it is in CONUS.

CHAPTER 3

Deployment

Ranger units can quickly deploy by any means of transportation to any location in the world.

3-1. Readiness.

One ranger battalion is always in an advanced readiness condition as the Ranger Ready Force (RRF) available for immediate worldwide deployment. A second battalion is prepared to deploy later and a third battalion to follow. The regimental headquarters maintains command and control, liaison, communications, and reconnaissance elements immediately available for deployment. Higher status of readiness in response to specific world situations can be achieved.

3-2. Ranger ready force.

In order to maintain advanced readiness conditions, ranger battalions are rotated as the RRF.

a. Before taking over the duties of the RRF, battalions must do the following:

(1) Complete preparation for oversea movement (POM) qualification, to include an individual records check and required inoculations.

(2) Test fire all weapons, confirm battlesight settings, and repair or replace faulty weapons.

(3) Assemble and load on pallets the unit basic load (Classes I and V) and other supplies and equipment.

(4) Check unit-derived packing list components for accountability and serviceability.

(5) Prepare special operations troop listings.

b. During an RRF period, ranger units:

(1) Have a positive recall system for all personnel.

(2) Store enough air items for a battalion-size parachute assault.

(3) Conduct no off-post exercises, unless approved by the regimental commander.

c. Each unit keeps an alert roster and a telephone notification system for recalling personnel. The rosters ensure that the duty officer and the unit charge of quarters make a minimum number of calls. Each commander briefs personnel assigned to his unit on the recall system and tests it often.

d. Ranger units can maintain an increased readiness condition for long periods, but they should not be kept in that condition longer than needed. The effort to maintain this level of readiness will degrade the unit's training and effectiveness. If the needs of the operations dictate long periods at the highest readiness condition, provisions must be made for sustainment training. Also, added support will be required from the RSE.

e. The ranger regiment maintains a strong family support system when the unit is training or deployed on combat operations. Each member of the chain of command is responsible for family support.

3-3. N-hour sequence.

The N-hour sequence is a departure time sequence designed to help the ranger force complete all required deployment actions within a certain time. The length of the N-hour sequence depends on the readiness condition of the battalion before deployment. The N-hour sequence begins when the battalion is alerted and ends when the first elements depart. The N-hour sequence does not include travel time to the objective area.

3-4. Intelligence and security.

a. **Essential elements of friendly information.** Items of sensitive information and subject areas, if not secured, can be collectively analyzed by a hostile force. This analysis can reveal the intentions of a unit or compromise a mission. These items of information are designated as essential elements of friendly information (EEFI). They are never transmitted without encryption. The EEFI associated with a ranger unit alert normally are the mission, the use of staging bases, nature of the target, special terrain or weather conditions, and time limits. The ranger force commander must ensure that EEFI are not discussed outside of secure areas or made available to personnel without a security clearance and a need to know.

b. **Security plans.** Security plans for billets, marshalling areas, and out-loading areas must remain flexible to adapt to any situation. Certain items are always addressed in security plans. These are access control and physical security.

(1) **Access control.** Ranger units control access to their respective areas by allowing only authorized personnel to enter or leave. All entrances are secured and an armed guard with the access roster is posted. Access to telephones is restricted.

(2) **Physical security.**

(a) Equipment is open to theft and sabotage during the early phases of deployment. Units must provide guards to secure all unit and personal equipment.

(b) Weapons, ammunition, and sensitive items must be stored in a restricted area and guarded.

(c) Classified documents are open to compromise during deployment. Steps must be set up for maintenance, transportation, turn-in, and destruction of classified material. The most secure facility available must be used.

c. **Rear detachment security.** Security after deployment (maintained by the rear detachment) is as important as security before deployment. No information should be given to the news media from other than official sources. The rear detachment may continue to play a role in the unit cover and deception plan.

d. **Tactical operations center security.** Only personnel listed on TOC access rosters are allowed free entry into the TOC area. Personnel wanting to enter the TOC, who are not listed on the access rosters, are detained until their identity and purpose are proven. Such personnel are escorted at all times while within the TOC area.

e. **Marshalling and out-loading area security.** The ranger unit normally marshalls its personnel and equipment in the unit area, providing its own security force. The USAF security police provide this support at the departure airfield. On Army airfields, Army military police provide security. En route security is provided by the RSE.

f. **Personnel and unit equipment sanitization.**

(1) Before deployment, personnel are relieved of non-mission-essential material having intelligence value. This reduces the intelligence available to hostile forces if they capture a ranger force member.

(2) The ranger force commander sets the degree of sanitization IAW mission requirements.

(3) If strict sanitization is imposed, guidelines are issued as follows:

(a) The only personal items retained are identification card and tags.

(b) Unit insignia and markings are removed from uniforms and equipment.

(c) Personal items, such as wallets and photographs, are secured by the rear detachment commander.

(d) Personnel are briefed on how to compose and retain combat diaries. No personal diaries are allowed.

(4) The commander decides how much US currency is to be taken with the unit. '

g. **Area sanitization.** To avoid the compromise of classified or sensitive mission-related material, searches are conducted before and after leaving the host installation, REMAB, or ISB. These searches are held to locate and remove any material, indicators, litter, or other items that could compromise the ranger unit's mission. Unit commanders ensure that a thorough sanitization of their areas is conducted, to include dismantling rehearsal areas, destroying terrain models, erasing chalkboards, and emptying all classified and nonclassified waste containers. During their searches, commanders must pay close attention to emergency operations centers, tactical operations centers, and marshalling and loading areas. The ranger unit's intelligence officer coordinates support from local counterintelligence elements to search the area again after the ranger unit leaves.

3-5. Operations.

a. **Actions upon receipt of the OPORD.**

(1) Task-organize and equip units for special mission(s) IAW the warning order.

(2) Allocate/reallocate mission-essential equipment.

(3) Distribute augmentee personnel.

(4) Redistribute key personnel,

(5) Prepare a deployment time sequence.

(6) Prepare a movement plan.

(7) Prepare and maintain a current operation estimate.

b. **Advance elements.** A planning element is normally deployed to the REMAB/ISB or JTF headquarters in advance of the main body if the ranger force does not deploy directly to the objective area.

c. **Emergency operations center.** An EOC is setup by the ranger battalion headquarters. The EOC is responsible for relaying all reports to higher headquarters. It remains operational until ordered to cease.

d. **Air movement.** The ranger force always deploys by USAF aircraft. This is normally an administrative air movement to a REMAB or ISB. It may be followed by tactical air deployment from there. For some missions, tactical deployment begins at the unit's home station. Even if the movement is administrative, key personnel are cross-loaded. This is done under the direction of the operations section air officer based on aircraft allocations and load plans.

3-6. Logistics.

The following describes logistics readiness needs and steps for deployment.

a. **Supply.**

(1) **Class I.**

(a) Combat rations are normally palletized when the unit is RRF. These rations are to be used during the tactical operation. If the mission calls for deployment directly into the objective area, the palletized stock of rations is off-loaded in the unit area and issued as needed.

(b) Each ranger deploys with emergency combat rations.

(2) **Classes II and IV.**

(a) **Individual clothing and equipment.**

- Individual A bags are packed IAW a predetermined packing list, tagged, and marked. They are palletized, moved to the departure airfield by the RSE, and loaded IAW the air movement plan.
- Individual B bags are packed, assembled, and forwarded on order by the rear detachment.

- Individual environmental equipment is issued. Equipment shortages are replenished out of stocks maintained by the RSE.

(b) **Air items.**

- While the unit is on RRF, a contingency load of main and reserve parachutes is maintained by the RSE.
- Individual and unit air items are palletized by the RSE and shipped in bulk to the REMAB or ISB.
- For direct-deployment missions, air items and packing material are taken to the unit by the RSE.

(c) **General supplies.** The RSE maintains pallets of common contingency items and special equipment. Pallets are picked up by the out-loading support unit and moved to the unit area for issue or moved with the other pallets to the departure airfield.

(d) **Bulk supplies.** Barrier material is identified, relocated, palletized, and out-loaded IAW the air movement plan.

(3) **Class V.**

(a) The unit basic load of ammunition is secured, palletized, and loaded by the RSE IAW the air movement plan.

(b) For direct deployment missions, selected pallets are taken to the unit area and ammunition is issued. The convoy route is coordinated through the post EOC, and secured by the RSE.

(c) Designated items are rigged by the RSE for airdrop, if enough time exists before movement to the REMAB or ISB begins.

(4) **Class VI.** Personnel deploy with comfort items as needed.

(5) **Class VII.**

(a) All weapons and sensitive items not deployed are assembled and secured by the rear detachment personnel.

(b) Unit vehicles are prepared for airland or airdrop.

(c) Inoperable mission-essential equipment is repaired or replaced.

(d) Special equipment requests are forwarded to the logistics section as they are identified.

(6) **Class VIII.**

(a) Individuals deploy with one set of individual Class VIII items.

(b) Unit medics deploy with modified aidbags.

(c) Controlled medical items (less drugs) are palletized and deployed with section supplies. Controlled drugs are secured and issued to unit aidmen on order.

(d) Medical supply chests are maintained in the unit aid station.

(7) **Class IX.**

(a) Units deploy with initial prescribed load list (PLL) items.

(b) Critical PLL shortages are identified to the RSE for follow-on issue.

b. **Maintenance.**

(1) Upon alert, priority maintenance or replacement of key items of equipment is coordinated by the logistics officer. Only minimum mission-essential equipment is replaced.

(2) Items to be returned from supporting maintenance activities are identified, and pickup is coordinated by the property book officer.

c. **Out-loading coordination and support.**

(1) Upon alert, the RSE provides a liaison officer to the ranger logistics section.

(2) The support platoon leader (SPI,) attends the air movement conference and coordinates the following:

(a) Movement times (for both troops and equipment) to the departure airfield.

(b) Steps for weighing and joint inspection of the pallets and vehicles upon arrival at the departure airfield control group (DACG).

(c) Limitations on the pallet height and weight due to aircraft type, allowable cargo load (ACL), and availability.

(d) Changes in the planned number of pallets due to the mission requirements or airframe availability.

(e) Rigging and loading of the heavy drop and supply bundles.

(3) Empty pallets are placed in position for loading by the RSE in coordination with the SPL.

(a) Companies and staff sections prepare pallets.

(b) For airdrop operations, vehicles are prerigged at the unit location. On order, vehicles are driven to the heavy drop rigging site. The SPL and the officer in charge of the rigger detachment coordinate for a rigging site and the delivery of the required air items to that location. Rigging is done by the RSE, and supervised by a member of the ranger unit. Loading aboard the aircraft is IAW the air movement plan and supervised by RSE and USAF aerial port personnel.

(4) The DACG is provided by the host installation. Upon arrival at the DACG area, unit personnel and equipment are under the control of the DACG. As the deployment begins, a

member of the ranger unit is present at the DACG to ensure an orderly loading of personnel and equipment.

(5) The DACG plans for troop-holding areas in the event of delayed or aborted aircraft.

(6) Unit movement data are maintained by the air officer in the operations section.

3-7. Communications-electronics.

a. **Alert notification communications.** Alert notification of the ranger battalions by the regiment is done by secure communications means. Units in local training areas are informed by radio or through their assigned range field telephone. Units operating out of normal FM radio range during RRF periods set up a high frequency net, or other communications, to their higher headquarters.

b. **Telephone service.** Strict control is maintained over key telephones being used during unit alert. Care is taken to ensure that COMSEC and OPSEC violations do not occur. All telephones are assumed to be monitored; classified, sensitive, or official information is not discussed.

c. **Radio communications.**

(1) **Voice communications.**

(a) A secure radio net is set up early in the deployment. The ranger unit EOC, the installation EOC, and the RSE EOC operate in this net. A liaison officer from the ranger unit is placed at the installation EOC.

(b) The ranger support element liaison officer, attached to the logistics section, operates a secure voice station on a separate logistics net to the RSE EOC and the RSE commander.

(2) **Cryptographic material.** During RRF, COMSEC codes and enabling equipment are kept available to set up the secure voice net.

3-8. Remote marshalling base and intermediate staging base.

a. Ranger units can deploy from their CONUS base directly to the objective area. A more common method would be for the ranger unit to deploy first to either a REMAB or an ISB before insertion into the area of operations. (See Figure 3-1.)

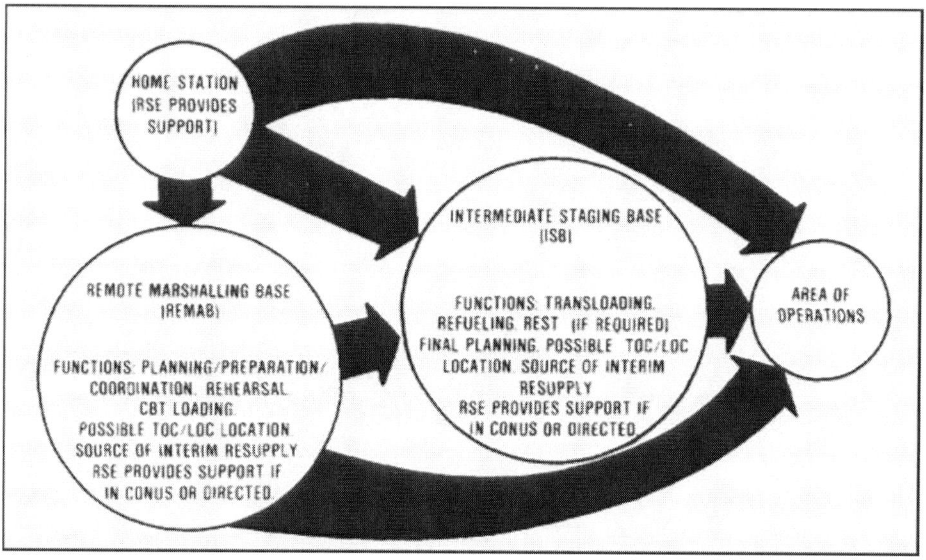

Figure 3-1. Deploying ranger units.

b. The REMAB is a secure base to which the entire ranger unit, including organic and attached support elements, deploys and where mission planning continues. The REMAB is normally within the geographical area encompassed by the command authority of the theater or JTF commander. This ensures that the CSS elements providing support to the ranger unit are operating within their normal area. It prevents or lessens out-of-sector support requirements for CSS elements.

c. The REMAB also provides a secure location for the ranger unit to conduct detailed planning and coordination with the controlling headquarters staff. In the REMAB, the ranger force commander conducts rehearsals, refines and modifies plans, determines PIR, and coordinates with the proper intelligence source to receive that information. It is in the REMAB that specialist augmentees to the ranger force are integrated into the unit, if they have not already joined. Specially trained supporting units, such as aviation and communications elements, also join the ranger force at the REMAB. The REMAB must provide--

(1) Access to the controlling headquarters staff.

(2) Physical security of billeting, planning, maintenance, and communications areas.

(3) Mess, billeting, latrine, and shower facilities for the ranger force and its supporting elements.

(4) Access to a C-141 or C-130 capable airfield, possibly with all-weather operations.

(5) Access to secure communications and processed mission intelligence.

(6) Access to rehearsal areas where sites can be built and live-fire rehearsals can be conducted.

(7) Access to the unit locations of major supporting elements, such as naval landing craft or Army aviation units.

(8) An external security force and an active counterintelligence agency.

(9) Vehicular transport for troop lift, equipment transfer, and administrative use.

(10) Access to maintenance support facilities.

(11) Medical support facilities to augment the ranger force medical personnel.

(12) Covered area to allow for packing of parachutes and rigging of airdrop loads.

d. The ISB is a more austere location. Elements of the ranger force deploy to an ISB to perform final planning, coordination, and task organization. The unit's organization and composition are finalized for movement to the objective area. The ISB is normally not occupied for long periods. Ranger units may deploy to the ISB from the REMAB or CONUS. Deployment to the ISB is more common when terrain or distance precludes insertion to the objective area directly from the REMAB or CONUS. Quick-reaction missions often involve the use of an ISB. Facilities needed at the ISB are limited to the following:

(1) Austere messing arrangements for the combat elements of the ranger force and its attachments.

(2) Austere billets or rest areas.

(3) Water point.

(4) Fuel for aircraft and vehicles.

(5) Areas for test firing of weapons.

(6) Covered and concealed areas for assembly of the ranger force and rigging of parachutes and door bundles.

(7) Austere airfield support facilities, possibly capable of all-weather operations.

(8) Remote location far away from civilians or traffic routes.

(9) Security and counterintelligence elements.

(10) Secure communications.

e. The REMAB and ISB should be located in an area similar in terrain and climate to the objective area. Time spent at the REMAB or ISB lets the ranger force begin acclimatization. Ranger units train yearly in all climates and emphasize physical conditioning. This allows acclimatization quicker than units not as well trained or conditioned. However, ranger units still need an acclimatization period if the terrain and climate of the objective area are different from that of their CONUS base. Not enough acclimatization means that the ranger force could be less efficient, especially when the objective area is hot, dry, or at a great altitude.

CHAPTER 4

Insertion, Extraction, Escape, and Evasion

Insertion is the entrance of a ranger force into a hostile area to execute a particular mission. It is the first critical phase of a ranger operation. These operations often involve deep penetrations of enemy territory by air, water, or land. This chapter emphasizes fundamentals and techniques employed by ranger units when their mission requires insertion and extraction. Extraction is conducted quickly after the mission to avoid casualties.

4-1. Intelligence.

a. Insertion plans are based on timely and accurate intelligence. The headquarters directing the operation must provide current intelligence information to the ranger force. The ranger regiment normally takes action to solicit required information from the controlling headquarters and from other agencies, if security permits. The type of intelligence needed varies depending on the specific mission, but normally includes the following:

(1) Current photographs from aerial or overhead reconnaissance.

(2) Detailed maps.

(3) Data on the population, terrain, and weather from area studies.

(4) Current enemy order of battle, enemy troop locations, and enemy capabilities.

(5) Details on the target area and infiltration routes.

b. To meet its need for current intelligence, the ranger force also relies on aerial reconnaissance and surveillance of the terrain over which it will move. Emphasis is on getting information on the enemy's ability to detect and engage the inserting forces. The location and abilities of air defense radar and weapons systems are critical. Terrain analysis is detailed and focuses on areas suitable for the insertion. Weather information must be detailed and current.

c. In getting information and intelligence data, OPSEC measures are stringent and must be enforced. Coordination with other agencies is kept within security constraints approved by the directing headquarters.

4-2. Deception.

There are always plans to deny the enemy knowledge of the ranger unit's insertion, or to deceive him as to the location or intent of the operation. Feints, false insertions, and other deception operations all add to ranger deception plans. Selection of unexpected means, time, place, and routes of insertion, as well as use of speed and mobility, give less reaction time to the enemy. Planning may include the use of diversionary fires to direct the enemy's attention away from the inserting unit. Deception techniques that may be used include--

a. Multiple airdrops, water landings, or both. This precludes the entire force from becoming engaged at the same time and place, if detected.

b. Dispersion of insertion craft (air or water) both in time and location.

c, Landing a force in an area closer to other potential targets than to the actual targets.

d. Leaks of false information and false messages.

e. False insertions or radio traffic to disguise actual insertions.

f. Diversionary actions such as airstrikes in other than the target areas.

g. Increased reconnaissance flights, either by manned aircraft or remotely piloted vehicles, over false targets.

4-3. Speed and Mobility.

a. Speed of action must be emphasized when the enemy can react quickly to the ranger force presence or when there are limited means of withdrawal. The operation must be conducted swiftly, allowing the ranger force to withdraw before the enemy can react.

b. Major problems facing ranger units are the need for speed and mobility, and the heavy loads carried by soldiers for sustained operations. Ranger units are required to be flexible and to be able to react to any problem within the objective area. However, if equipment is carried to cover every encounter, the load will become too heavy. This will reduce the speed and mobility of the ranger unit. The ranger force commander must always have a balance between the equipment carried and the mobility of his unit. He does this by making three different types of loads: fighting, mission, and existence. He decides what items will make up these loads, where they are carried, and what will be done with them upon contact with the enemy.

(1) The fighting load consists of those items of equipment, weapons, ammunition, food, and water that are common to all soldiers and are necessary for immediate use in combat. These items are normally carried and are retained at all times while in contact with the enemy.

(2) The mission load consists of those items of equipment or ammunition that are required by the mission, but are not normally carried by all rangers in a unit. The load may include spare batteries, demolitions, radios, antitank weapons, or mines. This load is normally spread out among members of the unit and is rotated often. It may be dropped upon enemy contact, but is normally carried to the objective and either used or emplaced there.

(3) The existence load consists of any items designed to provide protection from the elements: sleeping gear, changes of clothes, spare rations, or tentage. This load may or may not be brought into the objective area, depending on the tactical situation. If it is brought, it is normally dropped upon enemy contact or left in an assembly area and retrieved later.

4-4. Stealth.

Methods and steps chosen for insertion must emphasize stealth and try to avoid detection by the enemy. Multiple routes or methods of insertion may be used to preserve the ranger force if some soldiers or units are seen.

4-5. Suppression.

Effort must be made to suppress enemy detection devices, weapons systems, and command and control facilities. This may include electronic countermeasures or direct attack. This helps prevent the enemy from finding and engaging the ranger force en route to the objective area. Deception tactics add to suppression activities. If the ranger force is inserting by infiltration, suppression must not disclose friendly activities.

4-6. Security.

Emphasis on OPSEC prevents compromise of an impending operation. The operation must not be disclosed by overt rehearsals or training activities, the open use or procurement of special items of equipment, or the location of the marshalling area. Several measures can help in maintaining security:

a. Guarded and secure facilities for the headquarters and staff during planning.

b. Issue of operation orders only after units are moved to secure marshalling areas.

c. Alerts, rehearsals, and training to get the local people used to conditions that will exist during actual marshalling.

d. A secure facility set up to isolate units selected for an impending operation.

4-7. Night vision and electronic devices.

a. Rangers use night vision devices to detect and avoid the enemy's forces and his detection devices. These devices also assist in controlling and speeding movement and allow units to traverse almost impassable terrain.

b. The use of electronic homing devices permits precise navigation during reduced visibility. These devices may also be used with visual marking systems.

c. Passive night vision devices are used to assemble rapidly and to reorganize.

d. The use of active night vision devices in assembling, and the degree to which drop zones or assembly areas are "sterilized," depends on the following:

(1) Mission success may depend on not being seen during insertion. When a clandestine insertion is made, the time needed to get to the objective will normally be greater, the use of passive devices will prevail, and the sterilization of the area will be vital. Each ranger must be briefed on sterilization plans and techniques for erasing signs of the insertion.

(2) When detection is likely and the mission depends on getting to the objective before the

enemy can react, speed becomes crucial to the success of the insertion. Insertion may be near or on the objective. Rangers must get in and out quickly. They do minimum sterilization of the area and can use active RSTA devices.

4-8. Rehearsals.

Rehearsals should be close to the actual conditions of insertions and extractions. Security of rehearsals and rehearsal sites is important to the success of the operation. Rehearsals are held on terrain and structures like those in the objective area. If such structures are not available, they are built by the ranger force and the RSE with its supporting engineers.

4-9. Sand tables and terrain models.

Sand tables are useful to acquaint personnel with insertion sites and surrounding terrain. The use of sand tables and terrain models during the issuance of prejump orders and briefings enhances an orderly and speedy tactical assembly.

4-10. Communications.

Reporting during insertion and extraction missions is normally by exception, using an operations schedule.

4-11. Planning insertions.

a. **Centralized planning.** Planning for insertion of ranger units is centralized at the lowest level responsible for the overall operation. This is normally at the joint task force level, with detailed coordination required among all supporting forces. These forces may include other Army elements, USAF, United States Navy (USN), United States Marine Corps (USMC), other US agencies, and allies. Preparation must include detailed contingency planning for use of firepower, shock effect, speed to break contact (if detected), control measures, procedures, and planning. When planning takes a long time, intelligence must be continually updated through all available sources. This includes overhead photography and ground sources to ensure that the initial planning data are still current. Planning stresses the use of deception. Escape and evasion plans, external communication nets, and the location of caches must be known to leaders at all levels.

b. **Reverse planning sequence.** Insertion planning is done in the reverse planning sequence with the ground tactical plan being prepared first. Actions at the objective serve as the basis for the landing and assault plan, the plan for insertion and movement, the loading plan, and the plan for marshalling. All plans are interrelated. Some may be developed at the same time. Consideration of the factors of METT-T guides all planning. Special considerations that apply to these operations are:

(1) **Mission.** The nature of actions at the objective and requirements for special items of equipment and nonorganic personnel impacts on how the unit inserts. When the mission requires rapid deployment, the fastest method of insertion may be needed. In other operations, where success of the mission depends on maintaining secrecy, speed may be less important.

(2) **Enemy situation.** Enemy disposition, strength, and security measures along the route(s) near the objective area affect the method of insertion to be used. Planners must take into consideration whether the enemy:

(a) Uses RSTA detection devices or not.

(b) Has a strong air defense system or possesses air superiority.

(c) Has gaps between his defensive positions.

(3) **Terrain and weather.**

(a) **Topography.** Terrain such as mountains, swamps, or heavy forests favors land infiltration. Deserts, arctic regions, and open prairies favor air or water insertion. Areas not inhabited or with a friendly but scattered population favor land infiltration. A heavily populated area, or one which contains large enemy forces, favors air or water insertion. Hydrography--tides, currents, surf, reefs, and sandbars--must be considered when planning water insertion.

(b) **Weather.** Bad weather with reduced visibility and high winds favors land infiltration, and adversely affects air or water insertion. Limited visibility favors land infiltration; good visibility does not. Adverse weather aerial delivery systems (AWADS) lessens the impact of reduced visibility as a limiting factor on air infiltration. High winds may preclude airborne insertion.

(4) **Troops.** The number of men to be inserted and their level of training may be limiting factors. A need for special skills may call for use of nonranger unit personnel whose physical stamina and ability to perform many infiltration techniques may not equal that of the rangers. Whether supporting forces and special items of equipment are available must also be considered.

(5) **Time.** The main concern is whether the ranger operation is quick-response or deliberate-response. During a quick response mission, planning, preparation, and rehearsal time is short. The timing of the insertion is crucial with respect to weather, the enemy, and if an insertion craft is available.

4-12. Air insertion.

The most rapid form of insertion is by air. Rangers and equipment can be delivered by parachute (static-line or freefall technique), fixed-wing aircraft (airlanded), or helicopter (airlanded, parachute, or rappeling).

a. Suppression of enemy air defense along the insertion corridor may be necessary. The USAF will normally assume the major responsibility for suppressing enemy air defense artillery (ADA) capabilities that may interfere with the inserting forces. This is done by a variety of sophisticated electronic countermeasures applied against enemy equipment and by strikes against known enemy positions. Army fire support elements, special electronic mission aircraft (SEMA), and ground jammers may participate, as well as Navy aircraft, jammers, and gunfire.

(1) A main danger area is the forward edge of the battle area (FEBA) or frontier areas where

the enemy employs most of his sophisticated weapons systems and air defenses. Other dangerous areas are population or troop concentration areas behind enemy lines protected by point air defense systems.

(2) Since the majority of the enemy's detection devices and air defense weapons may be located near the point of entry, fire support, smoke screens (even at night), and suppressive measures may be critical. The use of special items of equipment may be needed to counter the enemy's RSTA effort.

(3) Long-range fires (artillery and naval gunfire) should be planned on known and suspected enemy antiaircraft locations and on prominent terrain features along the route(,s). Once beyond the FEBA, inserting units may be beyond the range of conventional indirect fire. More reliance will then be placed on air assets for fire support and electronic countermeasures.

b. There must be close coordination between the ranger unit and the transporting unit. What the transporting unit does during the insertion has a great impact on the success of the mission. To decrease the chance of detection, maximum use is made of periods of reduced visibility and tactical cover and deception. Drop zones and landing zones are chosen behind tree lines, in small clearings in forests, or on other types of terrain not easily seen.

c. All flight routes over enemy territory should be planned in order to avoid occupied areas and enemy air defenses. Routes should complement the cover and deception plan. They should avoid giving away final ranger objectives.

d. In-flight emergencies, mainly in deep-pedetration flights, should be considered, The ranger force receives a preflight briefing on the route to be flown and is informed of checkpoints along the route. Simple ground assembly plans for contingencies are set up before enplaning. In an emergency, the leader of the unit involved decides whether to continue the mission or abort. This is based on METT-T factors, contingency plans, and distance to the objective as compared to the distance to friendly territory. Contingency provisions for air and water rescue are made.

e. Deciding the DZ location is a joint effort. After receiving input from the air and ground commanders, the joint task force commander makes the final decision. During air assault operations, the ground commander is responsible for the overall process. After reviewing MEYF-T factors, it may be determined that surprise is more important than speed, or vice versa. This may determine the number, size, and locations of DZs and LZs.

f. Planning and preparation continue during the marshalling phase of any operation using air delivery for insertion. A review must be made of items of equipment that are crucial to the ranger force reaching its target area, accomplishing the mission, and returning safely. The need for special supplies and equipment, their transport, the distance to be traversed, the delivery means, and external support are all factors that enhance or reduce the chance for success. Once the means of insertion has been decided, logistics planning continues with emphasis on requesting and coordinating external support. The following must be considered:

(1) Detailed planning for mission logistical needs.

(2) Extra sets of key items of equipment to ensure redundancy and self-sufficiency of

inserting units.

(3) Plans for accompanying, planned, and emergency resupply. The ranger force should prepackage loads to meet specific contingencies. It should consider the use of supplies and equipment already in the operational area.

(4) In quick-response operations, only manportable items of equipment needed to accomplish the mission are carried. Plans and signals must be established for resupply of key items.

(5) Specific plans to evacuate casualties.

(6) All attached personnel and augmentation equipment must be properly prepared for movement.

(7) Provisions must be made for evacuating captured enemy equipment and personnel.

(8) All inserting units should be cross-loaded. Items to be considered in a cross-loading plan are the ground tactical plan, the assembly plan, the configuration of the DZ, the light route and formation of the aircraft, and the number of aircraft available. Cross-loading ensures that key leaders and equipment--for example, door bundles--are evenly distributed throughout the assaulting force. If one or more aircraft abort or are shot down, some key leaders and equipment will still arrive at the DZ. This permits mission accomplishment. Cross-loading aids rapid assembly on the DZ in support of the ground tactical plan. If a ranger is to assemble near the leading edge of a DZ, he is loaded on the aircraft to exit near the front of his stick.

(9) A bump plan is used if an aircraft should abort while still on the ground and there are no spare aircraft. The bump plan ensures that key leaders and equipment are not left behind when the main body takes off. If an aircraft aborts on the ground, key personnel move to another aircraft and replace nonessential personnel. This concept promotes success on the objective.

4-13. Airborne insertion.

a. During the planning phase of an airborne operation, there is no room for error in judgment. It can be planned and conducted well only if there is a constant exchange of information. All personnel involved must be informed of all changes in plans, times, and locations.

b. The reverse planning sequence is the key to the success of an airborne insertion. Most of the planning is based on the ground tactical plan, followed by the air movement plan to support it.

c. The USAF is responsible for flight planning, initial point (IP) selection, and crew procedures throughout the flight. The senior ranger in the aircraft must remain oriented and keep abreast of any last-minute changes. He must coordinate with the aircraft commander.

d. Most airborne insertions are made at either very high altitudes (HALO/HAHO) or very low. For combat operations, drop altitudes may be less than 500 feet above ground level (AGL), depending on the type of parachute used.

e. In airborne insertions, emphasis is on the use of special delivery or navigational techniques since normal insertion is during periods of limited visibility.

(1) **Adverse weather aerial delivery system.**

(a) A self-contained navigational system in some USAF C-130 aircraft is AWADS. The system allows the aircraft crew to fly a specific route to a DZ during darkness or bad weather. The system works with a Doppler radar that determines the ground speed of the aircraft, and an AWADS computer programmed with special flight route information (start point, route, checkpoints, wind speed and direction, and air speed). These two instruments show where the aircraft is in relation to the programmed checkpoints. This eliminates the need for airdrops under visual weather conditions.

(b) Only certain aircraft in formation are equipped with the actual AWADS device. Other aircraft are equipped with station-keeping equipment (SKE). This SKE consists of a computer that tells the aircrew where their aircraft is in relation to other aircraft in the formation.

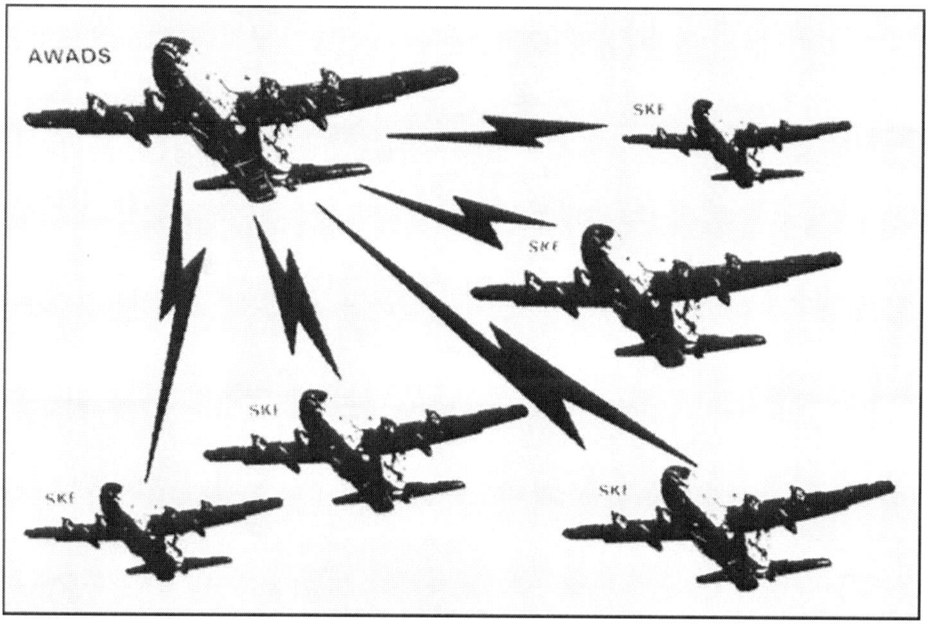

Figure 4-1. AWADS abd SKE aircraft formation.

(c) Airborne insertions using AWADS give the ranger commander the ability to airdrop personnel and equipment during low or zero visibility. This increases the safety of USAF aircraft. It protects them from visually tracked enemy antiaircraft guns and non-all-weather fighters. It also uses adverse weather as a combat force multiplier. Using AWADS, personnel and equipment may be inserted day or night without a pre-positioned USAF combat control team (CCT) or an Army assault team. Time length of the air train is shortened. The air corridor has to be cleared only once. This enhances the element of surprise. The CCTs are brought in with the assault elements. They are able to direct other low altitude parachute extraction systems (LAPES), container delivery systems (CDS), and air-land or heavy drop missions. Use of AWADS enables the ranger commander to conduct rapid, vertical reinforcement of units threatened by an enemy attack during bad weather.

(d) The use of AWADS requires detailed planning, rehearsal, and close coordination among the forces involved. Intense training of both USAF and Army personnel is required.

(e) Personnel must be briefed on terrain through use of photographs, maps, and terrain models. Limited visibility relies mainly on compasses and audible and visual assembly aids.

(f) During reduced visibility, essential heavy-drop serials may precede personnel aircraft. This reduces the time lag between personnel and heavy-drop chalks due to the need to move friendly troops off the DZ. Remaining heavy-drop serials or LAPES can be scheduled following P-hour.

(2) **Reconnaissance teams.**

(a) The operations of the main ranger force may be enhanced by use of reconnaissance teams. These teams are inserted ahead of the main body to provide early warning or to conduct reconnaissance of key objectives. If reconnaissance teams are required at a point along the flight path or near the main DZ, they may be cross-loaded on board any aircraft within the formation. They exit at planned times so as to land as close to their objective as possible. Should such teams be required at a point perpendicular to the axis of the main formation, one of the following is used:

- Another flank aircraft delivers teams to designated points before or with arrival of the main body.
- Selected aircraft deliver the main body, turn, and then deliver the reconnaissance teams.
- The teams are cross-loaded as complete elements on aircraft delivering other units whose flight path crosses the desired locations.

(b) Missions assigned to reconnaissance teams include:

- Providing local security for a USAF CCT, in which case, by joint doctrine, they form the Army assault team (AAT).
- Establishing surveillance or blocking positions on likely avenues of approach into the objective area or DZs.
- Locating and marking targets or assembly areas for the main assault force.

(c) Reconnaissance teams may be inserted by static line or HALO/HAHO operations.

(d) If the reconnaissance element is to locate a specific target or objective, it may be inserted up to 24 hours early. The longer the reconnaissance team remains near the objective area, the greater the danger of it being seen.

(e) In order for the information to be used by the ranger force commander, secure communications and a linkup plan must be prepared and rehearsed.

(3) **Assembly.**

(a) The ranger airborne force can conduct a quick assembly of combat power (weapons, personnel, and command and control elements) on the DZ. Quick assembly and reorganization is crucial because the unit is open to attack. Assembly must be fast and precise. Assembly areas and plans are set forth after considering all the factors of METT-T, mainly the location of the enemy, visibility, terrain, drop information, dispersion pattern, and cross-loading.

(b) The number of assembly areas depends on the location of the DZ, size of the assembly areas, and enemy's detection ability. Unit members move directly to the assembly areas without breaching security. Dissemination of information to all jumpers is the key to a successful assembly plan.

(c) For battalion-size or larger operations, general assembly areas are prescribed in the ranger force OPORD. Company commanders select and designate the exact location of their assembly areas. Although assembly area locations are terrain-dependent, they are generally on the DZ if the drop is at night and off the DZ if the drop is during daylight.

(d) Personnel chosen to carry assembly aids are cross-loaded in the center of their unit sticks. Strong, aggressive leaders should jump first and last in each stick. Upon landing, the leaders roll up their sticks toward the center, locate and move to the proper assembly aid. They may also carry or wear assembly aids. Individual rangers then follow these key personnel. During daylight operations, radiotelephone operators (RATELOS) can display color-coded pennants on their radio antennas to facilitate linkup with leaders. Key personnel may carry and jump with their own radios, and also display color-coded pennants on the radio antennas. Key personnel and RATELOs put their radios into operation before moving to assembly areas. For drops occurring near dawn or dusk, both daylight and night assembly plans are required.

(e) When tactical vehicles or door bundles are dropped, they are normally marked with panels, chemical lights, or strobe lights. The delivery parachutes are often color-coded.

(f) A ranger force should be able to assemble on an unfamiliar DZ, at night, within 45 minutes of the final aircraft pass. Because most ranger operations are time-sensitive, the unit normally moves from the assembly area to the objective when enough of the force has assembled to accomplish the mission. A straggler control point remains at the assembly area under the control of a designated leader. When everyone assembles, or at a specific time, the leader moves to link up with the unit. If available, a radio is left with the straggler control point.

(g) Assembly area security plans are prepared and disseminated. These plans must address hasty defense, sectors of responsibility, observation posts, fire support, and chain of command.

(4) **Contact on the DZ.** During the conduct of parachute insertion, rangers must be ready for enemy engagement at all times. Rangers are open to attack on the DZ before assembling. Just as contingency plans and chance immediate-action drills are developed, briefed, and rehearsed for other chance enemy contacts, rangers must be prepared to execute immediate-action drills for enemy contact on the DZ. The following actions are taken when there is enemy contact on the DZ:

(a) Rangers immediately assemble in fire team-size groups and fight off the DZ to designated assembly areas. Elements attempt to break contact and assemble.

(b) Assembly areas designated by commanders in the mission OPORD serve as primary rally points.

(c) Alternate assembly areas (normally 2 to 5 kilometers from the DZ) are designated in the mission operation order (OPORD). They are used by elements that cannot link up or assemble in the primary assembly area. These elements continue with the mission once assembled.

(d) During limited visibility, personnel should move directly and quickly off the DZ. They must avoid mistaking other friendly elements for enemy, which could cause firefights among friendly forces.

(5) **Reorganization.**

(a) The unit reorganizes according to plans, using designated assembly areas, aids, and identification markings for personnel and equipment. Replacements are named for key leaders or personnel who are lost or injured.

(b) Security teams assemble and move to their positions. Remaining units move quickly to their assembly areas, carrying supplies and equipment.

(c) Subordinate leaders keep the commander informed during assembly and reorganization.

(d) Runners may be used to report assembly status. This maintains radio silence. Designated personnel quickly recover supplies and equipment and move to the assembly areas.

(e) Reorganization is complete when all units have assembled or have been accounted for, and control has been established.

(6) **Pop-up technique.** This involves close terrain flying until the DZ is reached, then the aircraft quickly climbs to jump altitude. Special navigational equipment and aircrew training

are needed but may not always be available.

(7) **HALO/HAHO.** This involves high-altitude, low-opening and high-altitude, high-opening jumps with high performance parachutes designed to maneuver the jumpers to a specific point on the ground. During these operations, midair assembly procedures may be used.

(8) **Arctic.** Airborne insertion in extremely cold conditions is difficult. Times allocated for all parachute and equipment rigging must be increased. Rigging of individual equipment must be IAW FM 57-220. Ahkios can be rigged, packed inside a double A22 container, and dropped in platoon bundles. Because of the increased amount of individual equipment needed, allowable cargo loads for USAF aircraft are reduced.

(9) **Water.** Because of a lack of suitable land DZs, or to avoid detection, rangers may conduct airborne operations onto water DZs close to land. Procedures and techniques are similar to land DZ operations.

(10) **In-flight rigging.** On long-range operations involving extended flight time to the objective area, personnel may rig in flight. In-flight rigging requires detailed planning, coordination, of all jumpers. A rehearsal of these procedures should be conducted before deployment. Aircraft loads must be reduced to carry out in-flight rigging procedures. (See FM 57-220.)

4-14. Air-land insertion (fixed-wing).

a. **Planning factors.** Most of the planning factors that apply to airborne insertion operations also apply to air-land insertion operations. Most USAF aircraft cannot conduct air-land operations in total instrument weather conditions. The AWADS equipment is accurate enough for airdrop navigation, not air-land. Air-land operations are conducted, however, under conditions of extremely reduced visibility. Specially trained USAF crews, using sophisticated navigational techniques, can locate and land on small airfields in bad weather. As in airborne operations, ranger reconnaissance teams, as well as specially trained CCT elements, can jump before the air-land aircraft arrive. They emplace navigational and location markers to assist the pilots. Rangers may rely solely on air-land operations for insertion. This method does not provide for delivery of the force in the shortest time possible. Air-land insertions are normally in concert with an insertion of airborne elements.

b. **Time.** Since turnaround time is crucial, loads in aircraft should be unchained as the aircraft taxies. This also applies to personnel and seat belts. Aircraft should stop as close as possible to the assembly areas where personnel and equipment are off-loaded.

c. **Method.** Air-land insertions normally begin after a ranger Army assault team (AAT) and a USAF CCT have parachuted into the initial objective area and conducted reconnaissance, clearing, marking, and security operations. The ranger ground force commander is the arrival airfield commander. He has control over all arriving, taxiing, and departing aircraft (fixed-wing or rotary) through the CCT, which locates near him. The ranger force command and control element is usually on the first aircraft in and the last aircraft out.

d. **Security.** Air-land operations concentrate large numbers of men and equipment, which presents an open target to the enemy. The ranger ground force commander must ensure rapid off-loading, assembly, dispersion, and movement toward the objective.

e. **Communications.** The ranger ground force commander must have secure radio communications with the ground force, CCT, fire support elements, and higher headquarters. Wire should be laid to connect ground command and control locations.

f. **Operation duration.** Depending on the factors of METT-T, the aircraft may remain on the ground for extraction while the mission is being accomplished or take off at once. They may then orbit, pending recall, or return at a prearranged time.

g. **Extraction.** Extraction of the air-landed ranger force is normally planned and accomplished like an insertion, only in reverse order. A simple, yet effective, way must be found to see that no ranger personnel are left in the objective area.

4-15. Air assault insertion.

Ranger air assault (air-land, rappel, or combination) insertions normally take place during limited visibility. They quickly place the ranger force on, or close to, the objective. The ranger force (depending on the mission) normally plans on fighting when it hits the ground, completing the mission, and then being extracted. Air assault insertions need detailed planning, surprise, flexibility, speed, shock effect, concentration of combat power, and precise timing.

a. **Ground tactical plan.** The reverse planning process is important. The ground tactical plan, developed from the mission assessment, is the first planning area. All other planning begins from this point. The ranger commander and his staff decide stated and implied tasks, and request the proper helicopters for the mission.

b. **Fire support plan.** The fire support available to the ranger commander may be artillery, naval gunfire, mortars, attack helicopters, and USAF or USN tactical aircraft. Because of the timing of ranger air assault operations, preparatory fires are often not used. If they are used, enemy ADA sites normally have priority.

c. **Landing plan.** This plan puts the ranger force into the objective area at the time, place, and proper sequence to support the ground tactical plan. The ranger commander selects primary and alternate LZs and the priority for landing. The exact location of an LZ is determined after a detailed analysis of METT-T. Rehearsals are important. The ranger force must quickly assemble, reorganize, and accomplish the mission. The senior ranger on each aircraft ensures that the personnel on his aircraft offload at the correct LZ.

d. **Air movement plan.** The air movement plan provides for the coordinated movement of the ranger force into the zone of action in support of the 'landing plan. Flight routes, air movement tables, flight formation, altitude, and air speed should be considered. The ranger commander must have secure in-flight communications with his subordinate commanders.

e. **Loading plan.** This plan supports the air movement plan and provides for the pickup zone (PZ) setup and control. It sets priorities for the commander for loading personnel, equipment, supplies, and cross-loading. Air-loading table, aircraft loading formations, bump plan, and rehearsals should be considered.

f. **Staging plan.**

(1) This plan includes PZ selection and operation. The selection of PZs and LZs needs planning and coordination for best use of air assault assets. Site selection in or near the objective must be coordinated in person between the supported ranger commander and the aviation commander. The tactical situation is the key factor in PZ and LZ selection. Other factors are size of landing zones, surface conditions, ground slopes, approach and departure directions, prevailing winds, obstacles, communications, cross-loading, aircraft command and control, PZ and LZ markings, and rehearsals.

(2) The PZs are set up and secured about 15 minutes before touchdown of first aircraft. The fire support plan must support the other plans. All supporting fires must be coordinated with the air mission commander (AMC). Because of the depth of most objective locations, USAF tactical aircraft will be the main fire support method. Orbit locations for fire support aircraft must be planned.

g. **Other planning considerations.** These include escape and evasion, actions at the last LZ, assembly plan, downed aircraft procedures, control measures, forward area rearm and refuel point (FARP), weather delays, and deception plans.

4-16. Amphibious insertion.

Insertion by water may be by SCUBA, surface swimming, small boats, submarines, surface craft, helocasting, or a combination of these methods.

a. **Planning.** Planning for waterborne operations must be thorough. The plan is decided upon between the ranger commander, the transporting unit, and the joint headquarters directing the operation. Plans must be made for all types of enemy action and weather while on board the transporting craft. The transporting unit is given information on a need-to-know basis. Information that could endanger the operation may be withheld until the mission is under way. Initial planning includes:

(1) Time schedule of all events, from the beginning until the end of the joint operation, to be used as a planning guide. Timing for each event must be exact; the success of the operation depends on it. (See Figure 4-2)

Figure 4-2. Waterborne insertion.

(2) Embarkation site, where troops and equipment are loaded on the transporting craft.

(3) Drop site, where the ranger force will depart the primary craft and load into smaller boats.

(4) Handing site, where the unit will beach its boats, or land directly from amphibious craft.

(5) Loading, when loads are lashed and waterproofed IAW unit standing operating procedures (SOPs). Inspections are important. A bump plan for key personnel and equipment must be developed.

(6) Cross-loading, which will help preserve the ranger force. location of leaders, crew-served weapons, and needed equipment are stressed.

b. **Landing site.**

(1) **Beaches.** Beaches are checked to find one that allows an undetected approach. Landing sites that cannot be approached from different directions are avoided. Stretches of defended or patrolled coastline should be avoided. If sand beaches are used, tracks and other signs of

operation that may compromise the mission should be erased. Rural, isolated areas are best. The area behind the landing site should provide an assembly area and concealed routes for exit.

(2) **Barriers.** Reconnaissance may find that good landing sites are not suitable because of barriers between the site and the objective. Such barriers may be patrolled roads, large areas giving little or no concealment, enemy obstacles or defensive positions, areas populated with enemy sympathizers, or an area used for enemy troop maneuvers and training. Unless a means of avoiding such barriers is devised, landing sites should be limited to those without barriers.

(3) **Other factors.** Beach selection should also consider:

(a) The scheme of maneuver.

(b) Surprise.

(c) Enemy dispositions.

(d) Distance to the target.

(e) Landing and exit sites.

(f) Trafficability of the beach.

(g) Cover and concealment.

c. **Routes.** The route the transporting craft takes to the drop site should be planned to deceive the enemy. A major change in the route immediately after the unit's debarkation could compromise the mission. The route should be the same as a route that would be used in some other naval operation--for example, minelaying or sweeping, or a naval patrol. An alternate route must be planned.

d. **Tactical deception.** There should be other plans to deny the enemy knowledge of the insertion besides the water approach route plan. Plans may include electronic countermeasures (ECM), diversions, feints, ruses, or demonstrations.

e. **Actions at the drop site.** Some operations may permit landing directly from the transporting craft. If not, primary and alternate drop sites must be set. The drop site should be at least 1,500 meters offshore to cover any noise in the loading and launching. If the enemy has a surface radar capability, a drop site should be chosen to avoid detection. The drop site may have to be several miles offshore or may need the use of ECM. The manner in which men, small boats, and equipment are loaded must promote fast action at the drop site.

f. **Navigation.**

(1) Ship-to-shore navigation (to the landing site) may be done by dead reckoning. The course may also be maintained by compass navigation, reference to a shoreline silhouette, radar, or prearranged signals from the shore.

(2) Boat handlers, navigators, and other experts are relied on in various phases of the

transport operation.

(3) Lead reconnaissance teams or scout swimmer teams may be used to plant radar, radio, or infrared buoys, or distinct, shielded lights. Boats may be equipped with radar reflectors or beacons to permit accurate tracking.

g. **Actions on the beach.**

(1) Boats hold offshore while ranger scout swimmers move to a covered and concealed position. After reaching the beach, they make a short listening halt. They check the beach and the area immediately beyond the beach for enemy activity. Swimmers signal the force "all clear," mark the landing site, and provide local security.

(2) When the boats land, chosen personnel move directly into covered and concealed security positions. They prepare to defend the landing site.

(3) If the ranger unit is not carrying equipment, boats may be moved off the beach and buried or camouflaged. If heavy equipment makes it hard to move the boats, chosen personnel start unloading at once. The remainder of the force secures the landing site. Each boatload moves into position under the control of a designated leader. Teams are positioned with the best cover, concealment, and fields of fire available.

(4) The ranger force leader sets up a command post from which he can control the operation. He may establish secure communication with his headquarters.

(5) Boats may be deflated and buried or camouflaged near the landing site or away from it. If the boats are to be hidden near the landing site, a team may be chosen to dig holes or cut brush for camouflage along with the unloading of the boats. A small team may be chosen to sweep the beach to erase tracks and drag marks.

(6) The ranger force should be assembled and prepared to move within minutes after the last boat lands. Every ranger must be briefed on the unit's plan should enemy contact be made upon landing.

h. **Insertion by air from ship.**

(1) Helicopters from a ship can launch the ranger force from over the horizon and move at a low level to the LZ.

(2) Helicopters may be vectored from ships to a landing zone. The aircraft may return to the ship while the ranger force conducts the mission, they may orbit offshore, or they may stay at the LZ. Other aspects of landing and assembling are the same as discussed for helicopter insertions.

i. **Helocasting.** This form of insertion combines helicopters and small boats in the same operation. It is planned and conducted the same as air assault operations, except that the LZ is in the water. While helicopters are moving at low levels and low speeds, rangers drop small boats and jump into the water. They then assemble, climb into the boats, and move to shore. Ranger elements are usually assembled and loaded into the small boats within 20 minutes of the last ranger entering the water. This type of operation is normally conducted during limited visibility.

j. **Fire support.** Fire support may be ship-to-air, ship-to-shore, air-to-ground, or ground-to-ground. Specific missions include:

(1) Suppressing enemy air defense.

(2) Neutralizing enemy field artillery within range of boat or helicopter routes.

(3) Employing on-call fires against shoreline targets.

(4) Destroying enemy aircraft.

k. **Contingency planning.** The following must be planned for:

(1) Actions on enemy contact.

(a) En route.

(b) Helocast site.

(c) Flares.

(d) Aerial attack.

(e) Small arms fire.

(f) Indirect fire.

(2) Bump plan with priority to key leaders, crew-served weapons, and radios.

(3) Downed craft procedures.

(4) Escape and evasion.

(5) High surf.

(6) Adverse weather.

(7) Separation.

l. **Rehearsals.** The ranger force must rehearse all aspects of the amphibious insertion. This includes boat launching, paddling, beaching and assembly, boat commands, and capsize drills.

4-17. Land infiltration.

Land infiltration to the objective may be the best way to complete a mission. This is normal when the enemy has total air control or effective air defenses in the operational area to be penetrated, or when the objective is close to friendly territory. Rangers can infiltrate over any type of terrain, in any climate. However, thick forests, swamps, and broken or steep terrain offer the best chance for success.

a. **Planning.** There are many methods of infiltrating by land, and each has its good and bad points. Plans are designed to get the infiltrating force to the objective area without being seen. Plans should include:

(1) Selecting concealed routes for surprise and protection. Primary and alternate routes are based on detailed map and aerial photo study, ground reconnaissance, and data from agent reports.

(2) Avoiding obstacles, populated areas, silhouetting (when forced to cross obstacles or ridge lines), enemy positions, main avenues of approach, and movement along often used routes and trails.

(3) Selecting the time of infiltration to make use of reduced visibility, such as darkness, rain, snow, and fog, and periods of reduced enemy alertness, such as early morning, holidays, and times following protracted movement or combat. The time of the infiltration is important during critical phases (crossing borders or passing through enemy troop concentrations or populated areas).

(4) Using speed and stealth. All rangers must know routes, linkup points, alternate linkup points, time schedules, danger areas, and actions on enemy contact and at linkup points.

(5) Centralizing planning and decentralizing execution. Centralized coordination must ensure that units are acting IAW cover and deception plans. Control should be given to the leaders of the infiltrating groups. Within their designated zones, these leaders select their own routes and formations. Leaders control the rate of advance. When the enemy has electronic detection devices, radio communications must be kept to a minimum. Key leaders normally go with the group that is using the fastest lane or the lane least likely to be detected.

(6) Seeding infiltration routes to the assembly area with unattended ground sensors. These sensors can help the infiltrating force by reducing the chance of surprise contact during the infiltration, and by helping in controlling the forward progress of the units.

(7) Planning for fire support, which must include coverage of routes, danger areas, assembly areas, objectives, and PZs.

(8) Rehearsing actions at the objective, actions on enemy contact, actions at danger areas, and movement techniques.

b. **Movement.** Land infiltration is conducted using one of three techniques:

(1) **Movement in small groups along one axis.** All members of the force use the best route. Small groups are harder to detect, easier to control, and do not compromise the total force if detected. This technique may take too long and it needs an assembly area or linkup point. If the lead group is detected, other groups may be ambushed.

(2) **Movement in one group.** This technique has no reassembly problem. The same route is used by all, making navigation easier. A large force can fight out of a dangerous situation easier than a small one. It has no coordination problems between other infiltrating units. This form of movement is easier for the enemy to detect. If the force is detected, the mission may be endangered.

(3) **Movement in small groups along several routes at the same time.** This technique avoids putting the total force in danger, is less likely to be seen, forces the enemy to react in

many locations, and makes it harder for the enemy to determine the size of the force or its mission. Some groups travel over poorer routes and have problems of assembly, control, and support; the capture of personnel may endanger the mission. Regardless of which technique is used, ranger units normally move to reduce the chance of enemy contact.

c. **Assembly area.** Following movement into the general area of operations, infiltrating units converge on the assembly area or linkup point. Assembly may begin at once after the passage of lines, or it may take place at any time before reaching the objective. The infiltrating force must remain undetected before assembly. If an infiltrating unit is seen at this time, the enemy may learn the location of the assembly area, and the objective. It may take action to destroy the infiltrating force.

d. **Linkup point(s).**

(1) Selection characteristics include terrain that--

(a) Is free of enemy troops.

(b) Offers cover and concealment from enemy observation and patrols.

(c) Does not restrict movement of friendly units.

(d) Accommodates the unit.

(e) Is easily identified.

(f) Offers escape routes to alternate linkup points.

(g) Offers concealment from air attack.

(2) Action sequence for occupying the linkup point.

(a) Halt short of the point.

(b) Establish local security.

(c) Conduct a reconnaissance of the linkup point.

(d) Occupy the linkup point.

(e) Emplace observation posts and listening posts.

(f) Link up with other groups.

(g) Conduct final preparations.

(h) Continue the mission.

(3) Security may be enhanced with the aid of devices such as night vision goggles and night sights. Unattended ground sensors may be placed on likely avenues of approach into the linkup point. Normally, only passive observation devices are used to assist in avoiding detection.

(4) Plans must set a time schedule for actions at the linkup point. All unit members must move in quickly, consistent with security, so that the unit may proceed to the objective rally point (ORP) and on to the objective.

4-18. Stay-behind technique.

A ranger force can be positioned behind enemy lines by employing the stay-behind technique. In this method, a ranger force allows itself to be bypassed as the enemy advances. This is done to allow the ranger force to perform a specific mission. Stay-behind operations may require the concealment, or cache, of supplies before the enemy bypass. Rangers may need to use captured enemy supplies. The location of the assembly area and the plan for extraction are important parts of stay-behind plans.

4-19. Extraction.

The ranger force is normally extracted after the mission is completed. The enemy may also force the unit to abort its mission and to call for extraction. Contingency plans must be made to extract a force under enemy pressure, as well as one that is not in contact. Primary and alternate extraction sites are planned and coordinated with all ,porting forces. Ranger operations are normally conducted against deep targets. Extraction may begin with land movement but normally it includes the use of boats or aircraft. Plans may include primary I alternate means of extraction.

a. **Terrain.**

(1) The terrain is important in choosing the point of extraction - for example, whether or not the extraction site offers good cover from enemy direct-fire weapons and positions from which the extracting force can suppress the enemy. Tidal conditions and PZ size must be considered.

(2) Unlikely locations such as swamps, jungles, or mountain areas are often used for extraction. Specially trained air and naval crews, using the latest extraction devices, make such terrain useable.

b. **Enemy.**

(1) Extraction is best conducted without enemy pressure. However, such pressure could develop during the operation. Detailed plans are made for emergency or contingency extractions forced by the enemy as well as clandestine extractions not under enemy pressure. If under pressure, units will have to use planned fire support to suppress enemy weapons. At the beginning of the extraction with security posted to provide early warning, the unit may start to thin out the same as with a night withdrawal. Ground security may be relied on until the last unit withdraws under cover of suppressive fires.

(2) If enemy contact is made during an extraction, the commander must decide whether to reinsert units or break contact and move to another extraction point. Fire support and suppression are important and include suppressing enemy air defense by deception, ruses, or decoys.

(3) The ranger unit may have special weapons systems or equipment to aid in the extraction. Shoulder-fired antiaircraft missiles may be used to defend against enemy aircraft threatening

the extraction.

c. **Extraction by air or water.** Extraction by air or water is preferred when the resources are available and their use will not endanger the mission. Other factors that favor their use are when--

(1) Long distances must be covered.

(2) Time of return is essential.

(3) The enemy does not have air and naval superiority.

(4) Heavily populated hostile areas block land movement.

(5) Ranger forces are burdened with prisoners of war,casualties, or critical items of equipment.

d. **Extraction by land exfiltration.** Stealth and evasiveness are key elements of exfiltration. This method of extraction is favored when--

(1) The enemy has air superiority.

(2) The enemy is able to prohibit air or water extraction.

(3) The distance to friendly lines is short.

(4) The terrain provides cover and concealment (for movement on foot) and limits enemy mobile units.

(a) Multiple routes of exfiltration may be used to prevent destruction of the force if soldiers or units are detected. Captured enemy vehicles and equipment may be used to assist in the exfiltration.

(b) The ranger force may exfiltrate in one body or in small groups. Exfiltrating in small groups avoids the time delay of assembly.

(5) The exfiltrating force is lightly equipped and not burdened with captured personnel or material.

(6) The exfiltration route passes through an area occupied by friendly civilians or guerrilla forces who can assist the movement.

(7) Areas along exfiltration routes are not inhabited.

(8) The enemy force is dispersed or is not able to concentrate against the exfiltrating force.

4-20. Escape and evasion.

a. A vital part of all premission planning is the development of en-route and post mission E&E plans. Such plans enhance survival of the ranger force, and the transporting air or boat crews. The ranger force commander is responsible for--

(1) Helping in the development of an E&E plan in coordination with all supporting agencies.

(2) Ensuring that all members of the ranger force and the supporting boat or air crew are briefed on the E&E plan.

b. Each E&E plan is unique. Each situation has unique problems associated with E&E. The E&E plan devised by the ranger force commander must address these problems, while profiting from the individual abilities and training of the rangers and their supporting air and boat crews. The following considerations apply to E&E plans devised during ranger operations:

(1) The plans enhance survival of personnel who can no longer accomplish their assigned mission. The senior ranger in a group must decide if there are any remaining missions that the group can accomplish. If not, then he must try to evade and escape enemy capture if unable to link up and be extracted with the rest of the ranger force.

(2) Because of the depth of penetration behind enemy lines, most successful E&E plans may involve either air or water movement away from enemy-held territory. The air and boat crews supporting ranger operations are highly skilled and dedicated. If there is any chance of a successful rescue, these forces continue to try to retrieve rangers escaping from an enemy-held area.

(3) Rangers possess stamina and endurance. They can move a great distance (especially at night) and over rugged terrain to reach an area where they can rendezvous with search-and-rescue (SAR) aircraft or boats. Rangers are also resourceful and highly trained in communication, stealth, and camouflage. This, coupled with their self-discipline, means that successful E&E can be conducted over a great distance, through heavy concentrations of enemy, and over a long period.

(4) Escape and evasion plans for ranger elements should include avoiding contact with the local civilians. However, the aid of friendly insurgent forces may be enlisted. The ranger E&E plan may also include the use of other special operations forces E&E networks that are in place behind enemy lines. These nets must not be compromised by the volume of evaders.

c. Escape and evasion plans should be based on three phases:

PHASE ONE: That portion of the insertion following entry into the hostile zone where personnel survival and recovery is the only thing considered. Should an aircraft or boat carrying rangers be disabled during phase one, the ranking survivor assumes command. He must consider E&E planning and the experience and skills of other survivors. He should be concerned with the security of the mission, caring for wounded, and moving away from enemy activity while contacting SAR elements for recovery.

PHASE TWO: That portion of the insertion that is near the objective and could permit the ranger element to successfully pursue its mission. Although part of the E&E plan, actions by the ranger force during this phase become a forced land infiltration rather than an E&E. The senior surviving ranger must decide if enough personnel and equipment can be moved to the objective area in time to accomplish the mission. If so, he must proceed, taking along any surviving air or boat crew members. If not, he may proceed directly to the extraction point and wait, or he may contact SAR elements for recovery.

PHASE THREE: That portion of the operation after mission accomplishment. It is a hard time to escape and evade since the enemy is alert and reacting to the ranger forces. The confusion caused by the violence of the attack, coupled with the rangers' ability to move quickly cross-country away from the objective area, offers the best chance of a successful escape and evasion. The surviving ranger elements that cannot be extracted can do one of the following:

(1) Move over land at once to a PZ or beach for extraction by SAR elements.

(2) Move to an area where reentry into friendly lines can be made.

(3) Move to a neutral or nonaligned country where either internment or return to friendly control can be effected.

d. Any plan for E&E should provide for the issue to each ranger of an area map, a covert means of recognition, and a small compass. An emergency ration may also be included and any items needed for the climate in that objective area. The ranger force commander must balance the need for other survival equipment against the added weight and bulk involved.

CHAPTER 5

Strike Operations

The ranger regiment conducts strike operations in support of the Air Land Battle. Ranger strike operations are integrated into an overall plan designed to destroy, delay, and disorganize the enemy. They may also cause him to divert a major part of his combat forces and power to rear area security. Strike operations also help create an environment in which other special operations forces can operate. Strike operations can complement air power in disrupting enemy forces in depth. They allow the theater commander to regain the initiative and lead to decisive action.

5-1. Application.

a. The ranger regiment can conduct strike operations either as deliberate-response or quick-reaction missions. It can be inserted into the area of operations by land, sea, or air.

b. Strike operations can be conducted throughout the depth of the battlefield in support of conventional or unconventional warfare. Multiple strike operations can have a synergistic effect and can create situations that help friendly forces seize and maintain the initiative.

c. Strike operations include raids, personnel and equipment recovery, and interdiction of lines of communication. They are best accomplished by ranger units who use conventional techniques and special skills. Strike operations vary in size. When well augmented and supported, the ranger regiment can conduct strike operations deep in enemy territory using up to three battalions. Strike operations normally are limited in time and may be conducted unilaterally or in conjunction with other special operations forces or allied forces. They are normally ended by the extraction or exfiltration of the ranger force.

5-2. Raid operations.

a. A raid is a strike operation conducted behind enemy lines against strategic objectives, targets of high tactical value, time-sensitive targets, or key personnel and facilities in enemy rear areas. It normally involves a swift penetration of hostile territory to confuse the enemy or destroy his installations. Ranger units can conduct deep penetration raids when given the appropriate assets for insertion.

b. A strategic raid by the ranger regiment would be directed by the NCA and conducted under the OPCOM of a unified or JTF commander. An operational raid would be directed by the unified or JTF commander and conducted by that command directly or under the OPCOM of a corps commander. Typical targets would include:

(1) Command, control, communications, and intelligence centers at front level or combined arms army level.

(2) Nuclear and chemical weapons storage sites and delivery means.

(3) Key logistical centers, warehouses, ammunition complexes, or fuel pumping centers supplying logistical support to a specific front.

(4) Air defense and air traffic control integrating centers and air defense weapons sites.

(5) Key power generating and transmitting facilities, hydroelectric dams, and irrigation systems.

(6) Key ports or rail complexes.

(7) Key installations or facilities, such as airstrips, buildings, bridges, dams, or tunnels.

c. The size of the ranger raid force is determined by a detailed analysis of METT-T. The unit committed to the ground phase of the raid mission is kept as small as possible. It may be as small as a ranger company or it could include all the battalions of the ranger regiment. The force must be tailored to complete the mission quickly, violently, and with few casualties. The size of the force may also be affected by the augmentation of specially trained personnel needed for that mission.

d. Raids are normally conducted in five phases. (See Figure 5-1).

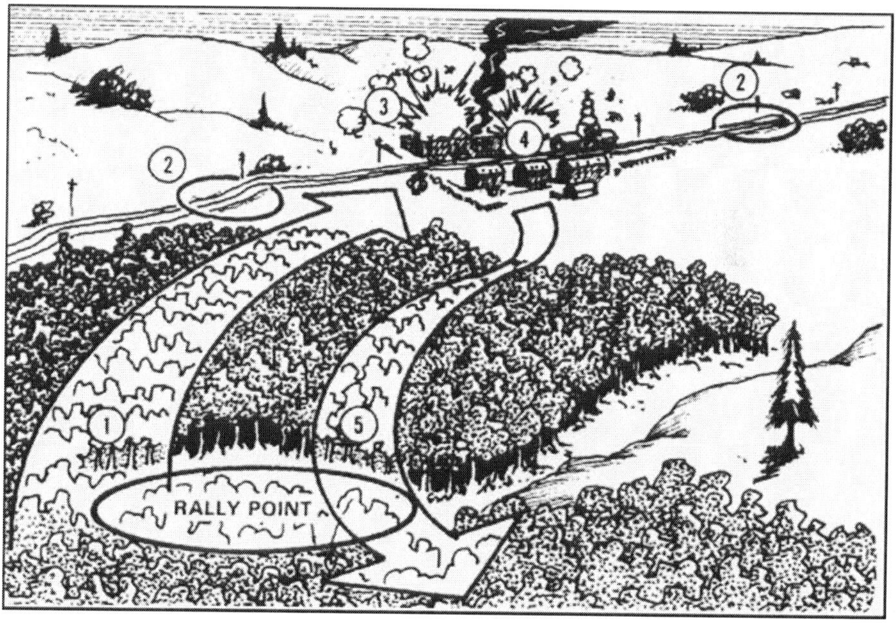

Figure 5-1. Five phases of a raid.

PHASE 1. The ranger force inserts or infiltrates into the objective area.

PHASE 2. The objective area is then sealed off from outside support or reinforcement, to include the enemy air threat.

PHASE 3. Any enemy force at or near the objective is overcome by surprise and violent attack, using all available firepower for shock effect.

PHASE 4. The mission is accomplished quickly before any surviving enemy can recover or be reinforced.

PHASE 5. The ranger force quickly withdraws from the objective area and is extracted.

e. This paragraph discusses the raid after insertion/infiltration. All considerations discussed in Chapter 4 must be applied in order for the unit to get deep behind enemy lines. Upon receipt of a raid mission, the following basics are emphasized:

(1) **Maximum use of intelligence information.** The gathering and dissemination of information must be continuous and provided to the raid force even while en route to the target area. To ensure mission accomplishment, the ranger unit must be kept informed of the latest enemy developments in the objective area to prevent being surprised. Only then can the full combat power of the rangers be concentrated at the decisive time and place.

(2) **Plan development.** The reverse planning sequence and the planning steps set forth in FM 101-5 are followed in the plan development.

(3) **Coordination.** Coordination is through the operational headquarters with friendly units (such as higher and supporting Army or joint headquarters, intelligence agencies) for fire, RSTA, special equipment, and personnel and logistical support.

(4) **Rehearsals.** Rehearsals help validate all aspects of planning for the raid and ensure precision in execution-They allow changes to be made in the plan before it is carried out.

f. A successful raid is ensured by--

(1) Launching the raid at an unexpected time or place by taking advantage of darkness and other periods of limited visibility, and moving over terrain that the enemy may consider impassable.

(2) Avoiding detection through proper movement techniques and skillful camouflage and concealment to include taking advantage of natural cover of the terrain, and using sophisticated equipment to detect and avoid enemy forces.

(3) Timing the operation as close as possible.

(4) Using all available support, both organic and nonorganic, to include use of special weapons such as Air Force "smart bombs" and artillery cannon-launched guided projectiles, with the ranger unit using laser target designators.

(5) Performing quick, violent, precise, and audacious actions that focus full combat power at the decisive time and place.

(6) Disengaging quickly upon mission completion.

(7) Withdrawing swiftly using planned routes and including a deception plan.

g. Four functions are normally performed by ranger tactical formations when conducting a raid. Each subelement is organized and equipped to do a specific part of the overall mission. Depending upon the specific mission, nature of the target, enemy situation, and terrain, these four functions are as follows:

(1) The command group controls movement to and actions at the objective. This unit normally consists of the ground commander, other subordinate leaders in the raid organization, and communications to support these leaders.

(2) The security element, whose organization is determined by the mission of the raid force, size and type of enemy force and its mobility and state of alert, terrain and avenues of approach into the area, and the time needed to seal off the objective area. The security element may perform the following:

(a) Secure the ORP.

(b) Give early warning of enemy approach.

(c) Block avenues of approach into the objective areas.

(d) Prevent enemy escape from the objective.

(e) Provide overmatch for the units at the objective and suppressive fires for their withdrawal.

(f) Provide short-range air defense.

(3) The support element provides the heavy volume of fire needed to neutralize the objective. Because fires from this unit are violent and devastating, they must be closely controlled to ensure the precision needed. On order or as planned, fires are lifted and shifted to cover the maneuver of the assault element by suppressing enemy fire from the objective or aerial fires. The support element may also be given specific locations to cover by fire in support of the security element if an enemy quick-reaction force moves toward the objective area. These may include routes into and out of the objective site, key terrain features, or installations adjacent to the main objective. Once the assault has been completed, or on order from the raid force commander, the support element displaces to the next planned position. Organization of the support element is determined by the following:

(a) Size of the objective, the geography of surrounding area, and the enemy threat (to include air) in the area. This element should be able to neutralize the objective (when supported by air or naval gunfire) and to lift or shift fires either when the assault is launched or when ordered by the raid force commander.

(b) Mission of the assault unit.

(e) Suitable firing positions.

(d) Size and nature of the enemy force in the objective area and those enemy forces capable of reinforcement at the objective.

(e) Fire support from other units (air strikes, naval gunfire, surface-to-surface missiles, and artillery fire).

(4) The assault element seizes and secures the objective and protects demolition teams, search teams, prisoner-of-way teams, or other teams.

(a) The organization of the assault element is always tailored to the mission. Each target area must be examined carefully. To assault, seize, and destroy an installation, position, or equipment, the assault element could be organized into one or more assault teams. The element's mission is to overcome resistance and secure the objective and to destroy the installation or equipment. Other specialized teams may also be needed. For example, sniper teams could be needed to remove key sentries. To capture prisoners, liberate personnel, and seize or destroy equipment, the assault element could be organized into assault teams, prisoner teams, search teams, medical teams, demolition teams, or breach teams.

(b) To destroy a point target or installation in a heavily defended area where the USAF cannot get close en to be effective, the assault element might be organized with one small team equipped with laser target designators. From covered and concealed positions, members of the team could then provide guidance for USAF delivery of laser-guided munitions from a safe distance.

h. The site chosen for the raid LZ or DZ must support the planned actions at the objective. There are two options when choosing sites:

(1) The ranger force can land on or near the objective and seize it before the enemy can react. (See Figure 5-2.) This avoids forced marches over land carrying heavy combat loads. If there is no suitable landing area near the objective, or the enemy has a strong reaction force nearby, this option is not favored.

Figure 5-2. Airborne landing on the objective.

(2) The ranger force can land unseen far from the objective. It then assembles, reorganizes, and moves into an ORP near the objective. (See Figure 5-3.) The objective is seized after security and support elements are in place. This option makes coordinated action easier by setting up control of small units before engaging the enemy.

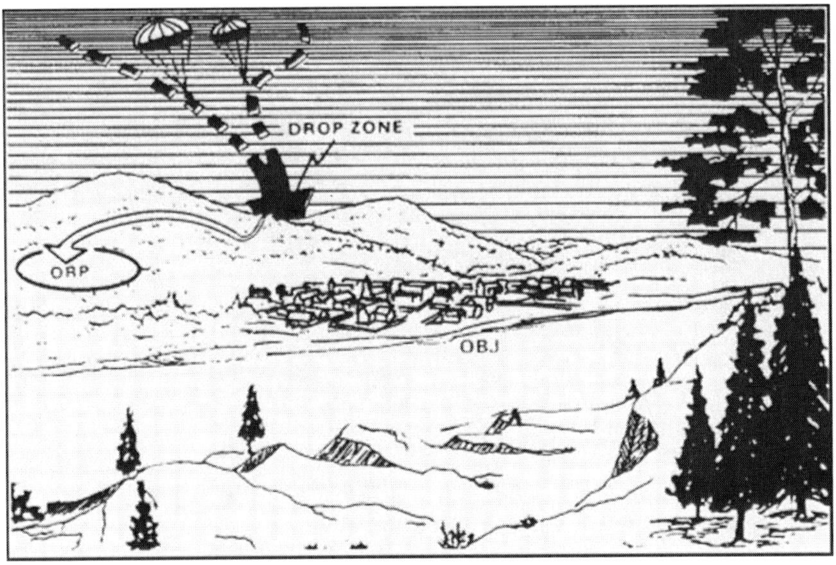

Figure 5-3. Airborne landing distant from the objective.

5-3. Interdiction operations.

a. AirLand Battle doctrine set forth in FM 100-5 is a shift in emphasis toward keeping the initiative in battle. The deep battle seeks out enemy elements separated from the main battle area and attacks them before they can be deployed. Interdiction of selected enemy forces by attacking them or blocking chokepoints hinders, delays, or interrupts the enemy's use of his lines of communication; it denies him the use of certain areas; it destroys industrial facilities, military installations, and equipment; and it stops or slows his advance along a particular route of approach. The ranger regiment's ability to conduct interdiction is vital to execution of the AirLand Battle doctrine.

b. Interdiction by ranger forces includes blocking or channeling approaches, holding key terrain for a limited time. and destroying facilities.

c. Ranger forces conduct interdiction missions that supplement those conducted by long-range Army and USAF weapon systems. Large-size ranger forces are normally used only when other systems fail or are not available, and the target is of such priority that it must be attacked. Small ranger laser target designator teams may work in concert with USAF or USN strike aircraft to direct terminally guided weapons systems. Rangers should be directed against crucial interdiction targets that disrupt, confuse, or delay the enemy and allow friendly forces to gain the initiative at the forward line of own troops (FLOT).

d. Typical interdiction operations are--

(1) Destroying rail lines, switches, rail yards, and rolling stock.

(2) Destroying bridges, overpasses, tunnels, and truck parks.

(3) Closing mountain passes or routes in restricted terrain.

(4) Cutting oil or natural gas pipelines and destroying pumping and distribution systems.

(5) Destroying dams, locks, spillways, and flood control gates.

(6) Destroying rail transload points, switches, and yard engines.

(7) Destroying shipyards, dry docks, wharfs, lighters, cranes, and stevedore facilities.

(8) Destroying electrical generation facilities, power lines, transformers, and grid monitoring centers.

(9) Destroying water pumping stations and filtration systems.

(10) Destroying radio and TV stations, phone lines, microwave terminals, satellite receiving stations, and telephone exchanges.

(11) Ambushing enemy forces and personnel moving along lines of communication.

e. Considerations when using rangers for interdiction are:

(1) Insertion and extraction.

(2) Support resources available.

(3) Ranger force vulnerability.

(4) Target identification.

(5) Threat environment.

(6) Other or supporting means of interdiction.

(7) Command and control.

(8) Time and magnitude of the effect on the enemy forces at the forward edge of the battle area.

(9) Collateral effects of target destruction.

f. Interdiction targets are not identified and attacked at random. They are part of an overall deep attack plan to destroy or cripple an entire system. Interdiction is based on the assigned deep campaign plan that directs which targets are to be attacked, by whom, the results desired, and the priorities of attack for specific systems. Based on his plan, the ranger force commander selects those ranger elements best suited to conduct the attack. Target selection, which requires detailed intelligence and planning, is based on:

(1) **Target criticality.** A target is critical when its destruction or damage would have a major influence on the enemy's ability to conduct or support operations. Each target is considered in relation to other elements of the particular target system designated for interdiction. The criticality of a target changes with the situation. For example, when the enemy has few locomotives, a railroad bridge is less critical.

(2) **Target accessibility.** The ranger force must be able to insert into the target area. Terrain, security controls, location, and means of insertion are considered when studying a target for accessibility. systems. Based on his plan, the ranger force commander selects those ranger elements best suited to

(3) **Target restorability.** The enemy's ability to restore a damaged facility to normal operating capacity affects target selection. The enemy's repair and maintenance ability and his reserve stocks must be considered.

(4) **Target vulnerability.** A target is vulnerable when it is open to attack by means available to ranger forces. Vulnerability is affected by the nature of the target such as type, size, disposition, and security. It is also affected by the means available to attack it such as explosives, incendiaries, and special devices.

g. Key to the interdiction process is the ability to locate and evaluate target systems for attack and the timing of the attack. A target system is a series of related functions that serve a common purpose.

(1) A target is one element, installation, or activity identified for attack such as a locomotive, train, bridge, or prison.

(2) A target complex is several targets in the same general area, such as a railway marshalling yard, an airfield, or port and dock facilities.

(3) A target system consists of an industrial system and its sources of raw materials; the rail, highway, waterway, or airway systems over which these materials are transported; the source of power and method of transmission; the factory complex; and the means by which the finished product is transported to the user. The targeting process is usually accomplished in the special operations force cell at EAC and the battle coordination element (BCE) at corps level.

h. Interdiction techniques include ambush, emplacement of conventional demolitions, target designation for laser-guided munitions, and emplacement of beacons or sensors.

NOTE: The enemy surface-to-air threat normally precludes the use of an airborne platform for direct observation and designation. This may be accomplished by the on-ground ranger force.

i. Through interdiction, ranger units are an ideal force to contribute to the Army's AirLand Battle operational concept. This includes taking the initiative, moving quickly to strike deep in the rear of the enemy, and defeating the enemy force beyond the FEBA. Planners and commanders must ensure that the ranger units are used on missions that are synchronized to contribute to the overall deep battle campaign. Strikes against critical targets must combine with efforts of other assets to take the initiative from the enemy, limiting his flexibility, initiative, and momentum.

5-4. Personnel and equipment recovery operations.

Recovery operations are specialized raids to liberate imprisoned or detained personnel or to return certain items of equipment to friendly control. These operations include recovering and extracting downed or hijacked aircrews and political or military leaders. Rangers use surprise and combat power to overwhelm resistance before harm can come to the detainees or prisoners.

a. **Planning.** The planning of personnel and equipment recovery operations is the same as for the raid. The main difference is that the ranger force commander must plan for the extraction of the recovered personnel and the loading and extraction of sensitive equipment. Personnel and equipment recovery raids are often executed as quick-response missions. They normally require the ranger force commander to plan and execute the recovery quickly.

b. **Augmentations.** This type of strike operation often requires that the ranger force be augmented with nonranger personnel with special skills or abilities. Examples are medical personnel, technical experts, mechanics or crew members trained in repair and retrieval operations, linguists or translators, and other special operations force elements.

c. **Organization.** The organization of the recovery force is the same as for a raid. Sometimes special teams are needed to perform certain missions involved in the recovery. This mission is conducted aggressively, usually ending in close combat during darkness or other limited visibility conditions. Reconnaissance teams may be inserted ahead of the ranger force to reconnoiter the objective and to locate the detained personnel.

(1) All planning and execution takes place as described for the raid. The emphasis must be on detailed, timely intelligence.

(2) Medical teams must be available to care for the detainees.

(3) The ranger personnel must be adept at searching and clearing of buildings.

(4) The ranger force must--

(a) Locate the specific building or area where the friendly detainees are being held.

(b) Gain entry into buildings, use stun grenades, and avoid obstacles and booby traps.

(c) Engage all enemy personnel using selective-kill techniques.

(d) Locate, identify, and secure detained personnel.

(e) Move personnel to a secured extraction site.

(f) Leave no detainees or rangers behind in the objective area.

d. **Execution.** The key to success is surprise and speed. The enemy must be assaulted when least expected. The ranger force must ensure that friendly detainees are not harmed during the assault - either by rangers or enemy personnel.

e. **Extraction.** The ranger force commander should plan for the extraction point to be as close as possible to the point of recovery. This prevents the detainees from having to move cross-country or for the ranger force to transport bulky equipment long distances. For small groups of detainees or small items of equipment, the extraction point can be further away. Recovery operations may use any method or combination of methods of insertion and extraction (see Chapter 4). Close planning and coordination is normally required with US Army, USAF, or USN aviation for both delivery and evacuation of the target area.

CHAPTER 6

Special Light Infantry Operations

Ranger units are normally employed on strike operations. However, broad US strategic needs demand a wide range of other military abilities. The ranger regiment may conduct special light infantry operations when larger conventional forces could not be used. The regiment conducts these operations using light infantry doctrine, methods, and techniques.

These operations include, but are not limited to, movement on foot, military operations on urbanized terrain (MOUT), passage of lines, ambush, and reconnaissance. Ranger units do not normally conduct planned operations against enemy armored forces. Battlefield surprises or METT-T considerations may cause ranger units to engage armored forces for short periods. The ranger regiment can conduct a conventional light infantry defense for short periods. A defense is not normally conducted adjacent to other friendly forces. It is normally a perimeter defense from which the unit then breaks out from encirclement and links up with friendly forces.

6-1. Movement.

a. Most missions involve movement on foot in tactical operations. Ranger units infiltrating behind enemy lines seek to avoid chance enemy contact. They choose the movement that allows them to retain security and control. To avoid loss of surprise and initiative, casualties, and mission failure, ranger units normally--

(1) Avoid chance enemy contact, if possible.

(2) Move on covered and concealed routes.

(3) Avoid likely ambush sites and other danger areas.

(4) Practice camouflage, noise, and light discipline.

(5) Maintain all-round security.

(6) Make unavoidable enemy contact with the smallest element possible.

(7) Retain the initiative to attack at the time and place of the ranger unit's choice.

b. In planning the movement of his unit, the ranger commander must consider all the elements of METT-T and determine how they will affect the movement. He must determine the route, formations, security measures, methods of command and control, location of key leaders and weapons, and action to be taken on chance enemy contact.

(1) **Routes.** The ranger force commander conducts a careful route analysis, using the factors of observation, concealment, obstacles, key terrain, and avenues of approach (OCOKA). He chooses a route that maintains the security of the force while ensuring surprise. He tries to choose a route that will avoid enemy contact and speed movement. The commander makes a detailed route reconnaissance using maps, aerial photographs, and reports from persons familiar with the area. If possible, he reconnoiters the route himself or uses a reconnaissance team. He may have that team act as guides or mark the route using covert marking devices.

(2) **Formations.** Ranger units use the movement techniques found in current light infantry doctrinal publications. The traveling, traveling overmatch, or bounding overmatch technique is chosen based on the likelihood of enemy contact. The ranger force normally tries to avoid detection and contact with the enemy while moving. The commander chooses a movement technique that ensures that if unexpected contact were made, it would be with the smallest element possible and not the entire formation. The distance between moving ranger elements depends on the terrain, visibility, and enemy situation. March interval is normally great enough to allow each succeeding element either to deploy or change direction if enemy contact is made. Distances are close enough that each element could quickly assist the element to its front.

(3) **Security.** Security is to the front, rear, and flanks when enemy contact is likely. A front security element is placed well forward with adequate communications with the main body. The security element is used to detect the enemy and warn the main body. The main body then either changes direction or hides until the enemy is past. If the enemy does detect the main body, the security element may be used to attack the enemy from the flanks or rear. Flank security elements are placed next to the route of march. They move forward either by alternate or successive bounds. An alternative is to have the security element moving adjacent to the column along routes paralleling the direction of march. Rear security is handled the same as front security. Plans are made for the rear guard to support the main body during chance enemy contact. This is done either by maneuver or by furnishing supporting fire.

(4) **Command, control, and communications.** All available communication is used, consistent with OPSEC and movement security, to assist in maintaining control during movement. March objectives, checkpoints, and phase lines may be used to aid in control. The number of reports is reduced, normally only exception reports are needed. The ranger force commander should be well forward in the formation but may move throughout as the situation demands. Communications with security elements are mandatory. Operations security often precludes the use of radios. Therefore, connecting files, runners, and visual signals are normally used. Detailed planning, briefing, rehearsals, and control are valuable if there is enemy contact. Alternate plans are made to cover all possible situations.

(5) **Weapons.** Rangers are placed within the formation to allow a mix of their individual weapons to cover the front, flanks, and rear of the formation. Rifles, grenade launchers, and light automatic weapons cover all sectors. Crew-served weapons such as machine guns, antitank weapons, and mortars should be given designated locations within the formation. This allows the element leader to employ them quickly. They should be placed near the lead element, but not so close as to be engaged along with it. Crew-served weapons are used to

provide overmatch, especially when employing the bounding overmatch technique of movement.

(6) **Rehearsals.** The ranger force, expecting movement as part of an operation, conducts the following combat drill rehearsals:

(a) Conduct movement.

(b) React to enemy contact.

(c) Maneuver under fire (fire and movement).

(d) Cross danger areas.

(e) Break contact.

(f) Conduct a hasty ambush.

(g) Make a temporary halt.

(h) Establish a perimeter.

(i) Establish a rally point.

(j) Take actions upon consolidation.

(k) Use arm-and-hand signals.

(l) React to flares and indirect fire.

c. After a ranger force is inside enemy territory, it must be alert to avoid detection before reaching the objective.

(1) Should the ranger force security element become aware of enemy presence, it alerts the main force. The main force tries to move away without alerting the enemy and continues the mission.

(2) If the enemy does make contact, the security element then engages the enemy, trying to deceive him as to the size of the ranger force and its direction of movement. After breaking contact, the security force divides into small groups and moves to a rally point to rejoin the main force.

(3) The ranger force does not fight en route to the objective unless there is no alternative, then it quickly breaks contact and moves away.

(4) Ranger units may need to set up a temporary defensive position for resupply, evacuation of wounded, planning, or extraction. One technique that can be used is the temporary perimeter defense.

(5) If the moving ranger force comes under indirect fire, it moves quickly out of the impact area and continues the mission.

(6) The most effective means of combating an ambush is through the immediate return of a

high volume of fire by all weapons. This requires alertness, discipline, dynamic leadership, and rehearsed plans.

(a) Regardless of the method of movement, all weapons should be positioned for immediate use. A well-executed ambush is hard to counteract. However, the effectiveness of an ambush can be reduced by the immediate return of fire. Enemy weapons may be silenced, thus creating weak points in the ambush position. Smoke and high-explosive (HE) grenades are effective in executing counterambush actions. They create confusion, provide a screen for movement, and disrupt the ambush plan.

(b) When all elements of a ranger force are not trapped within the ambush, the elements that are free to maneuver should initiate an immediate flank or rear assault against the ambush force. A flank assault permits better coordination with personnel in the ambush. Supporting fires are provided by those weapons that cannot be hand carried and fired in the assault.

(c) Specific actions against a near and far ambush are as follows:

- **Near ambush.** Elements caught in the kill zone return fire at once and move out of the kill zone. Elements not in the kill zone help by maneuvering and conducting a flanking attack. The unit assaults the ambush position.
- **Far ambush.** Those ranger elements not in the kill zone set up an overmatch position and lay down a base of fire for the elements in contact. The elements in contact move out of the kill zone and prepare to assault the enemy, using individual movement techniques and fire and movement.

(7) The reorganization after enemy contact involves the use of rally points, plans for local security, reorganization of the force, evacuation of dead and wounded, and movement based on the unit mission.

6-2. Ambush operations.

a. **Definition.** An ambush is a surprise attack, from a concealed position. It is used on a moving or temporarily halted enemy to destroy or capture him and his equipment. An ambush is a useful tactic because--

(1) Small, well-trained, disciplined forces, with limited weapons and equipment, can destroy much larger enemy forces.

(2) It reduces the enemy's overall combat effectiveness by destruction and harassment of his forces.

(3) Enemy morale and effectiveness suffer heavily at little cost to the force executing the ambush.

b. **Execution.** A successful ambush must be executed with precision, violence, speed, and audacity of execution. For success, ambush operations must emphasize the following:

(1) **Surprise.** Surprise, more than any other single aspect, enhances the value of an ambush. Surprise increases the potential for inflicting damage on the enemy with less danger to the

ranger force.

(2) **Coordinated firepower and shock effect.**

(a) Coordinated firepower is used for maximum shock effect. Massive volumes of accurate fire, explosives, and mines, coupled with an aggressive attack, breaks the enemy's spirit to fight back. Surprise increases shock effect and the chances for success. Shock effect can cover unexpected defects in an ambush--for example, ambushing a much larger force than expected.

(b) All weapons must be sited with interlocking fires in the kill zone and along likely avenues of entrance or exit. Mortars should be used if the terrain permits. Tripods and traversing and elevating mechanisms are normally used with machine guns to lock in fires. All riflemen use firing stakes to mark left and right limits, and elevation stakes. There is a tendency to shoot high in an ambush--especially at night. The M203 grenade launchers are sited to cover the dead space and routes of escape. When the rules of engagement permit, plastic bags containing CS crystals may be taped to the front side of Claymore mines.

(3) **Control.** Control is essential.

(a) Leaders must have contact with all members of their unit to alert them to the oncoming enemy. Leaders should not move around the ambush site during this crucial period. A method used to alert members can be to tie strings or vines to rangers' legs or arms. By a series of light tugs, all members of the ambush can be alerted to enemy presence.

(b) The leader must initiate the ambush with a casualty-producing device. A bank of Claymore mines on a double-ring main is an excellent device to spring an ambush. Other good techniques are to use a 90-mm recoilless rifle firing antipersonnel (APERS) rounds or a machine gun. All must be under the direct control of the ranger leader. Whistles or pyrotechnics must not be used. They will allow the enemy time to react.

- As soon as the enemy is hit, he reacts. The ambush force has only a few seconds to destroy the enemy in the kill zone. He will quickly recover from the initial shock and leave the kill zone--either with a direct counterattack or withdrawal, Subsequent fires and other banks of Claymore mines must be planned.
- The leader always executes the ambush, except when a member of the ambush knows he has been discovered. He then has the authority to execute--with killing fire, not by yelling.
- The cease-fire must be controlled by the ranger leader. A whistle or other device may be used to get attention, and then cease-fire is signaled.

(4) **Security.** The flanks and rear of an ambush site are open to counterattack. Flank and rear security may be enhanced by -

(a) Echeloning in depth.

(b) Designating sectors of observation.

(c) Positioning of RSTA devices.

(d) Enforcing noise and light discipline.

(e) Having a good withdrawal plan.

(f) Securing routes of withdrawal.

(g) Executing with speed and violence.

(h) Positioning security forces to sea] off the ambush area.

(i) Having good camouflage.

(5) **Simplicity.** A simple, direct plan improves the chance of success. The ambush plan must be clear yet concise to offer the greatest likelihood of success. For example:

(a) Mission statements for security, support, and assault elements must be clear, concise, and direct.

(b) Tasks to be performed by the ambush elements should be easy to understand.

(c) Contingency plans should be simple.

(d) Routes into positions and withdrawal routes should not cross. They should be the shortest, most secure routes.

(6) **Training and self-discipline.** Precise and violent execution is how ranger units win against enemy forces. All advantages must be exploited. Discipline must be strict. There must be no sleeping, talking, eating, or smoking in the ambush site. If an ambush is to be set up for long periods, then the elements of the ambush must be pulled back at set times to the ORP for rest. Extended ambushes of 24, 36, or 48 hours require six- or eight-hour shifts. It may take a company to man an extended platoon ambush position. Tired troops cannot man an ambush well. Other vigorous operations cannot be performed all day and have an alert ambush all night.

(7) **Maximum use of specialized equipment.** This includes the most sophisticated weapons, communications equipment, and RSTA devices. The effectiveness of any mission is enhanced by the use of this equipment.

(8) **METT-T.** Every ambush mission must be evaluated by the ranger commander IAW METT-T to design an ambush that works.

(a) **Mission.** Clearly state the mission--for example, stop enemy resupply, ambush enemy armor columns, or capture personnel or equipment.

(b) **Enemy.** State the type of enemy units--for example, guerrillas, main force, logistics carrying parties, or armor units. What is the size of the force, its habits when attacked, and type of support? What is the enemy night vision ability and expected time of movement?

(c) **Terrain.** Is it wooded, mountain, jungle, or urban? Some good ambushes can be set in a city or suburb. Are the routes available to the enemy suitable to his activity?

(d) **Troops.** Squads and platoons are the basic ranger ambush forces. An ambushing force should be able to engage a force two to three times its size--but evaluate enemy and terrain, A platoon ambushing an enemy tank company in daylight in open terrain is risky. But engaging that same column at night or in a forest may give the ranger platoon an advantage.

(e) **Time.** How long will it take to emplace the ambush? A complex, lethal ambush with plenty of "dirty tricks" takes hours to set up; a hasty ambush on a trail takes 15 to 30 minutes. How long is the ambush to be in place? What hours of the day, or at nighttime only? How soon must it be set up? Do not set up an ambush in darkness--even if it is to be executed in darkness. Try to allow every ranger to '6see" and "lay" in his weapon in daylight. Time is important.

c. **Organization of the ambush force.** A ranger unit conducting an ambush must be task-organized to perform the following functions: assault, security, and support. The ambush forces should be task-organized according to the TOE--by platoons, squads, and fire teams. The TOE should not be changed to create smaller elements for an ambush. The TOE formations may be reinforced with machine gun, recoilless rifle teams, or a 60-mm mortar squad.

(1) **Assault.** The elements assigned the assault mission either move directly into their positions or through a release point. The mission may include any combination of the following:

(a) Conduct the main assault.

(b) Halt an enemy's motorized column or any moving target,

(e) Kill or capture personnel.

(d) Recover supplies and equipment.

(e) Destroy vehicles and supplies.

Search teams are not always used. The ranger leader must decide how and when to use search teams. When rangers leave the security of their well-chosen, concealed ambush position, they are subject to the fires of the enemy who may also be hidden and ready. If it is at night, do not throw tripflares or shoot illumination to light the search area as this will illuminate and expose rangers to the enemy. Always assume there is hidden enemy--the ambush will not kill them all. Night observation devices (NODs) or a red filtered flashlight taped to M16s should be used to quickly search in ranger buddy teams. If the return fire from the enemy is great or if the ambush missed the main body, then the leader may choose to break contact and leave without searching the kill zone.

(2) **Security.** The elements assigned the mission of security may move to their positions

directly or by way of a release point. Their missions may include any or all of the following:

(a) Secure flanks, rear, or ORP.

(b) Provide early warning.

(c) Seal off the kill zone to prevent the enemy from escaping or reinforcing.

(d) Assist in executing the ambush.

(e) Cover withdrawal of main ambush force.

(3) **Support.** The units assigned a support mission provide fires that may include employment of--

- Heavy automatic weapon fires.
- Antitank fires.
- Mortar fires.
- Mines.
- Flame munitions.

d. **Ambush site.**

(1) **Choose an ambush site.** When choosing an ambush site, all sources of information must be used to enhance surprise, exploit the enemy's weak points, and take advantage of the terrain. An analysis of the terrain is also based on the other factors of METT-T to include maps, aerial photos, and, if possible, personal reconnaissance. Emphasis is on--

(a) Natural cover and concealment for the ambush force.

(b) Routes of entry and withdrawal (at least two) that should be direct and easy to reach.

(c) Good observation and fields of fire.

(d) Harmless-looking terrain.

(e) Few enemy escape routes.

(f) Limited enemy reinforcement ability.

(g) Nearby assembly or rendezvous area.

(h) Terrain that will canalize enemy into kill zones, and natural obstacles to keep him there.

NOTE: Try to select a site covered by friendly supporting indirect fires.

(2) **Take advantage of the terrain.** Emphasis must be on exploiting all natural cover and concealment afforded by the terrain. Site the ambush and individual positions based on the terrain rather than trying to adapt the terrain to a fixed geometric design.

(3) **Restrict enemy movement.** Restricting enemy movement by natural or man-made

obstacles should also be planned.

e. **Types of ambushes.**

(1) Ambushes have two categories for ranger purposes--area ambush and point ambush.

(a) Area ambushes may be set up by ranger platoons, companies, or battalions. They are used to interdict enemy movement in a given area or inflict casualties on his forces. Area ambushes consist of a series of point ambushes. The size and location of the ambushes are dictated by the METT-T analysis.

(b) Point ambushes are set at the most ideal location to inflict damage on the enemy. Such ambushes must be able to accept the enemy force from more than one direction. The ranger force levels may enable it to execute an ambush in two or three main directions. The other directions must be covered by security that gives early warning of enemy attack.

(2) The basic ambush is linear (Figure 6-1). An ambush is set on an expected avenue of approach. This ambush can accept contact from three basic directions--left, right, and front. The rear is secured by a security team.

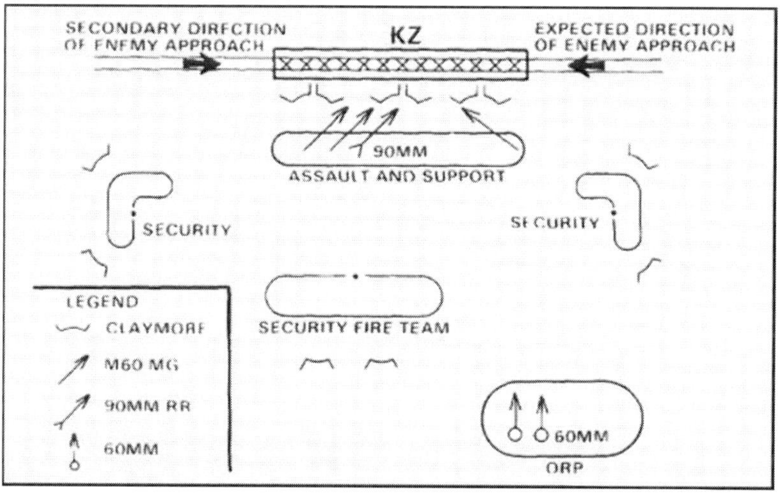

Figure 6-1. The linear ambush.

(3) A variation of the linear ambush is the L-shaped ambush (Figure 6-2). The L-shaped ambush is formed with the base (bottom) of the L perpendicular to the expected enemy direction of advance. This is a good ambush for a road, jungle trail, or an area where the enemy is canalized and his approach route is known. The L-shaped ambush can handle an enemy coming from the expected direction (toward the base of the L) from the front (stem

of the L). It is less effective against an enemy formation that comes from the opposite direction. The ambush must be executed when the enemy main body has cleared the base of the L. The flank security must protect the rear of the base of the L. In the L-shaped ambush, weapons must be sited to avoid direct or ricochet fire into friendly forces. The security elements must be considered as fire fans are being planned.

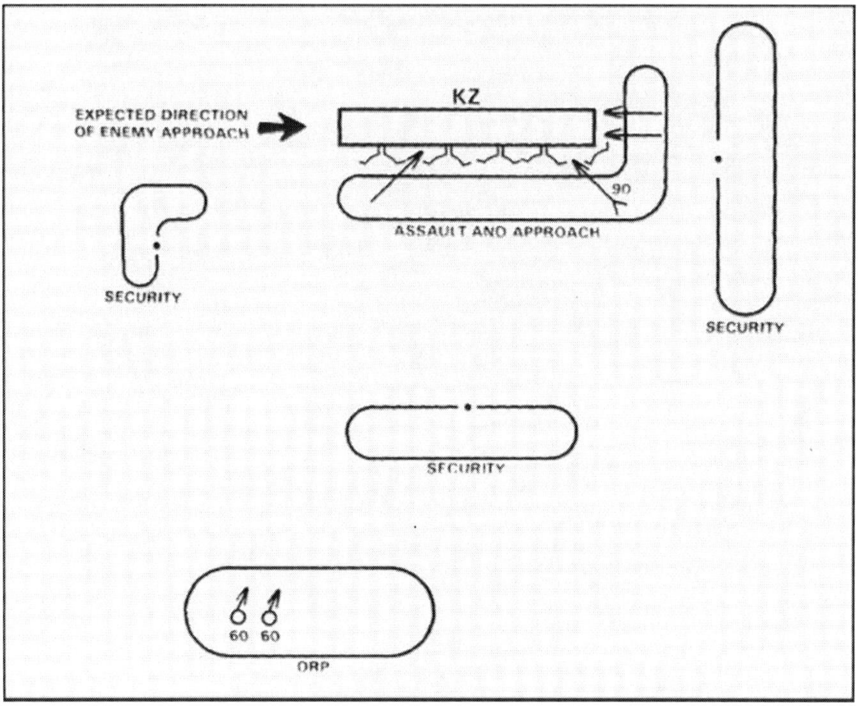

Figure 6-2. The L-shaped ambush.

(4) Other ambushes (the Z, the X, the V, and so on) are all variations of the linear and L formations (Figure 6-3). The more advanced ambush formations are difficult. Units must be expert in the basics before they are ready to employ such techniques.

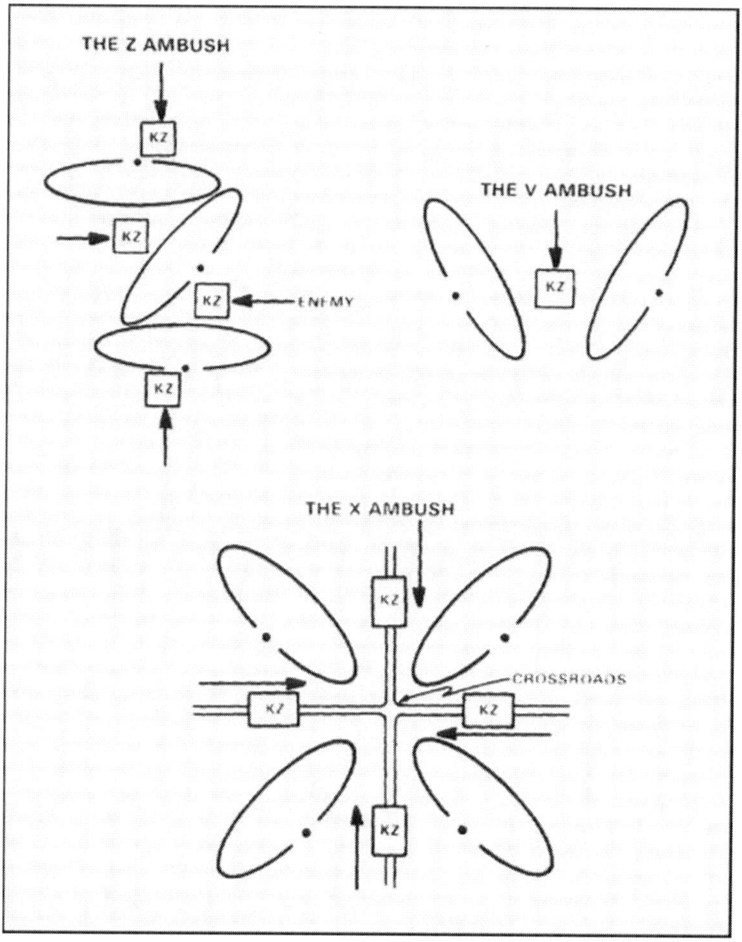

Figure 6-3. Other ambush variations.

(5) Another type of point ambush is the mechanical ambush (Figure 6-4). It consists of Claymore mines set in series with a double-ring main. It is command detonated or detonated by a triggering device activated by the enemy. Mechanical ambushes are normally manned. Rangers prepare to engage the enemy with direct fire if the mechanical ambush does not detonate or if it is wholly or partially ineffective. Mechanical ambushes are an effective way to interdict a large area using a small ranger force. If the mechanical ambush is effective and

rangers do not reveal their presence, the enemy is confused. This has a devastating effect on his morale and effectiveness.

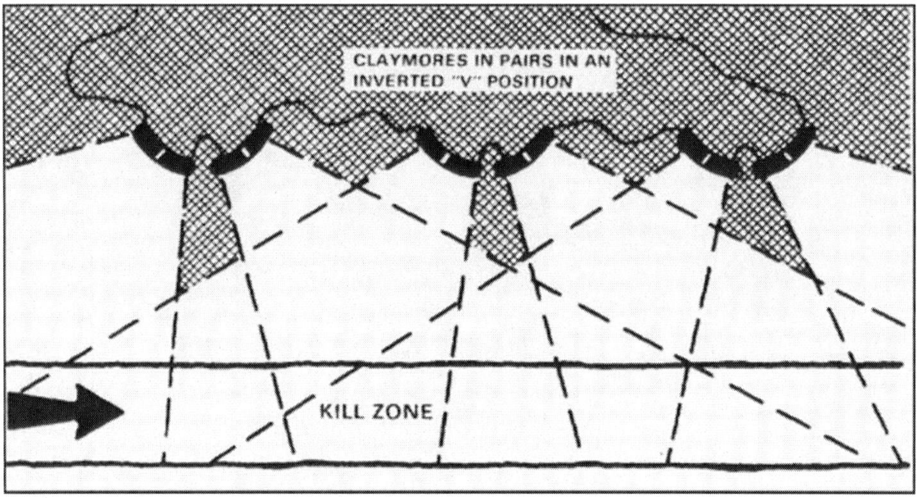

Figure 6-4. The mechanical ambush.

f. **Execution of the ambush.**

(1) **Occupation.** Stealth and security are important factors.

(a) Position security teams and early warning detection devices first.

(b) Use the rest route to main ambush position consistent with security.

(c) Quickly occupy the ambush position and set up communications and signaling devices.

(d) Position key weapons (automatic and antiarmor).

(e) Rig Claymore mines, tripflares, and booby traps.

(f) Ensure that all weapons are correctly positioned. Assign sectors of fire to provide mutual support and cover dead space.

(2) **Camouflage.** Each man must be hidden from the enemy, During preparation for the mission, each man camouflages himself and his equipment, and secures his equipment to prevent noise. At the ambush site, positions are prepared with minimum change in the natural appearance of the site. All debris resulting from preparation of positions is concealed to prevent evidence of occupation.

(3) **Movement.** Movement is kept to a minimum. The number of men moving at a time is closely controlled. Every man is as quiet as possible, especially at night. Light discipline is rigidly enforced at night. Smoking is forbidden.

(4) **Signals.** Audible and visual signals, such as whistles or pyrotechnics, must be changed often to avoid setting patterns and alerting the enemy. Three or four simple signals are needed to execute the ambush. Signals are used to--

(a) Provide early warning of an enemy approach. A signal by the security force to alert the patrol leader to the correct direction of enemy approach may be given. This includes arm-and-band signals, radio, or field telephone.

(b) Initiate the ambush. This may be the detonation of mines or explosives, Fire is then delivered at once in the heaviest, most accurate volume possible. Properly timed and delivered fires add to the achievement of surprise, as well as to the destruction of the target.

(c) Lift or shift fires, Signal for lift or shift fires, if the kill zone is to be assaulted. Voice commands, whistles, or pyrotechnics may be used. When the kill zone is assaulted, the lifting or shifting of fires must be as precise as when starting the ambush. Otherwise, the assault is delayed and the enemy has a chance to recover and react.

(d) Withdraw. A signal for withdrawal can be voice commands, whistles, or pyrotechnics.

(5) **Objective rally point.** The ORP is located far enough from the ambush site so that it will not be overrun if the enemy manages to attack the ambush unit(s). Existence loads are normally left in the ORP. Routes of withdrawal to the ORP are scouted (when possible by each man), Withdrawal routes should provide cover and concealment for the unit and hinder enemy pursuit. Withdrawal routes are a main consideration in the selection of the ambush site. They may be the key to survival after execution of the ambush. On signal, the ambush force quickly (but quietly) withdraws to the ORP. If pursued, the withdrawal may be by bounds, with mines or hasty ambushes used to delay pursuing forces.

(6) **Ambush variety.** If one method of ambush is used predominantly, the enemy will develop an effective defense against it and will be affected less by the shock of the ambush since he knows what to expect. More than one method of ambush should be used. No single method will fit all combinations of terrain, equipment, weather, and enemy capabilities. Variety should also be used when signaling - for example, use multiple signals, both audible and visual. Use weapons fire, mines, and RSTA when possible and vary signals to avoid compromise.

(7) **Swift action.** Speed in the execution of the ambush and the withdrawal should prevent enemy reaction forces from engaging the ambush force. Speed is often a shield against casualties and failure. When there is contact with reaction forces, speed may enhance quick disengagement. Ranger units must not only perform better than the enemy but also move faster than the enemy can react.

g. **Successful ambush.** Emphasize the following:

(1) Intelligence to ensure the enemy is ambushed at a time and place when he least expects or is least prepared to fight. This is preferred during reduced visibility to achieve surprise.

(2) Detailed planning and thorough training and rehearsing of all elements in all phases of the ambush. This ensures maximum shock effect through swift, precise execution.

(3) All available RSTA devices to permit daytime effectiveness at night when moving, shooting, or detecting enemy movement. To avoid detection, active RSTA devices should not be used until after the ambush has been triggered.

(4) All available firepower with emphasis on antiarmor, area and automatic weapons, and precision-guided munitions.

(5) Speed to achieve surprise and enhance security of the force.

(6) Cover, concealment, and overall protection afforded by the terrain when moving or when occupying ambush positions.

6-3. Antiarmor operations.

a. Antiarmor operations are not primary missions for ranger forces. They lack antiarmor firepower and have limited antiarmor ability. Ranger units can, however, perform limited antiarmor missions on close terrain by interdiction or ambush. The antiarmor weapons they do have are man-portable, but the ammunition is too heavy to carry many rounds cross-country. A ranger battalion has ten Dragon missile launchers and nine 90-mm recoilless rifles. It is also equipped with light antitank weapons (LAWs) and laser target designators (LTDs) that can guide Copperhead rounds.

b. Every ranger must be skilled with the limited antiarmor systems available. Proper selection, fortification, and camouflage of antiarmor ambush sites, route selections, withdrawal procedures, and initiative are important. Armor kill zones (AKZs) are normally set up after infiltrating the ranger force behind enemy lines under limited visibility.

c. A ranger antiarmor force is organized into four elements: command and control, support, security, and armor-killer teams (Dragons, LAWs, 90-mm recoilless rifles). These elements function the same as a raid force. The exception is that the assault force is the armor-killer team with the mission of destroying enemy armor, and it will probably not enter the AKZ. After the ambush occurs, the ranger force withdraws to its ORP. Indirect fires (if available) are then brought to bear on the AKZ.

d. Ranger units may conduct antiarmor ambushes and interdiction operation by use of laser-guided Copperhead rounds or air-delivered precision-guided munitions. Air-delivered family of scatterable mines (FASCAM) can be used to provide the ranger force security against attack by armored forces.

6-4. Military operations on urbanized terrain.

a. The ranger regiment normally tries to avoid the manpower-intensive, high-casualty battles associated with MOUT. However, due to the nature of strike operations and the types of targets selected, the regiment may conduct MOUT for limited periods. As in all ranger operations, the ranger force commander tries to surprise the enemy, attack him from unexpected directions, and destroy him with a violent, precise attack. The maze of cellars, rooms, sewer systems, rooftops, and alleys common to urban areas presents many opportunities for the ranger force.

b. Offensive operations by a ranger force exploit the centralized planning and decentralized execution aspects of ranger unit training. Individual action and initiative are always channeled to accomplish the commander's intent. The ranger unit's offensive tactics are based on quick action and boldness. Attacks by the ranger force are short and sharp. Execution is marked by swiftness, daring, timing, and surprise. The enemy's sleeping habits, eating habits, and relief times are watched. launched almost always at night or under smoke screens, attacks are often made without mortars or artillery, relying on surprise for success. Snipers and sharpshooters are used to shift the balance of battle.

c. Deliberate defensive operations are not good ranger unit missions. However, if a ranger unit must defend for a short period, the close terrain of urban areas allows it to function well against a stronger force. The buildings in a city break up advancing enemy formations and force them to move along streets. The ranger force executes an active defense anchored on strongpoints, such as a building or a group of buildings. It launches surprise counterattacks against the enemy's flank and rear. The aim of ranger defense is to give the enemy no respite and to make him feel threatened by a sudden, deadly attack. Ranger units should be withdrawn from defensive positions in urbanized areas as soon as possible. Prolonged MOUT and the associated casualties quickly degrade a ranger unit's combat effectiveness.

6-5. Perimeter defense.

The Perimeter defense may be conducted to control key terrain, gain time or economize forces.

a. **Planning.**

(1) **Warning order.** The order must be issued soon after mission receipt. It should contain the mission, time schedule (including rehearsal and inspection times, mission time, and place for OPORD), uniform and equipment, and special instructions.

(2) **Operation order.** The leader normally gives the OPORD on the ground to be defended. Locations and sectors of fire for squads and crew-served weapons should be designated. If time permits, leaders walk the terrain with subordinates as the OPORD is issued. The OPORD covers security, communications, signals, target reference points (TRPs), armor kill zones, and priority of engagement rules for antiarmor weapons. The OPORD also states the priority of work.

(a) Set up security and communications.

(b) Position crew-served weapons and personnel.

(c) Clear fields of fire.

(d) Prepare fighting positions.

b. **Organizing.**

(1) Squads are positioned with overlapping and interlocking sectors of fire and observation.

(2) The M60 machine guns are positioned either with an assigned final protective line (FPL) of fire or with a primary direction of fire (PDF). The PI)Fs are employed on covered avenues of approach such as gullies. The M60 FPLs should interlock across the unit front.

(3) The 90-mm recoilless rifles and Dragons are positioned with frontal cover to engage attacking armored vehicles from the flank.

(4) Riflemen are positioned to support and protect crew-served weapons. Riflemen are assigned FPLs that interlock at a point just beyond hand-grenade range.

(5) Dead space is covered with Claymore mines, antipersonnel mines, and M203 fires.

(6) Antitank mines, positioned to provide close-in protection from armored vehicles, are covered by LAW fires.

(7) Leaders plan indirect fire for long-range suppression of armored vehicles and short-range protection from dismounted assault. Mortar or artillery final protective fires (FPF) are placed on the most dangerous dismounted avenue of approach.

(8) Range cards are prepared for all crew-served weapons. Squad and platoon leaders prepare sector sketches and fire plans.

(9) The ranger force camouflages everything from ground and aerial observation.

(10) Leaders designate alternate and supplementary positions.

c. **Executing.** Leaders control the distribution and concentration of fires in their sectors. They constantly consider the overlap of fires with units to their left and right.

(1) Squads and crew-served weapons engage targets in sectors assigned by platoon leaders.

(2) Each weapon is used in its best role:

(a) Rifle, against enemy infantry within 300 meters.

(b) Machine gun, against enemy infantry within 1,000 meters.

(c) Grenade launcher, against infantry within 350 meters and armored vehicles within 200 meters.

(d) Light antitank weapon, against armored vehicles within 200 meters.

(e) Recoilless rifle, against armored vehicles within 400 meters.

(f) Dragon, against armored vehicles within 1,000 meters.

(3) Most dangerous targets are engaged first.

(4) Squads shift fires as needed or as directed by platoon leaders.

(5) The platoon masses fire on selected targets while maintaining coverage of the entire platoon sector.

(6) Platoon leaders control fires.

(7) After repelling an enemy assault, the platoon consolidates and reorganizes the defense.

6-6. Breakout from encirclement.

A breakout from encirclement is conducted when units, operating behind enemy lines, find themselves cut off from friendly forces and surrounded by superior forces. Given ranger unit missions, the chance of operating as an isolated force behind enemy lines is great. The breakout is characterized by determining enemy weak points, deception, massing of combat power, and a direct attack for a violent and timely breakout.

a. **Characteristics and fundamentals.** The breakout from encirclement involves the following characteristics and fundamentals:

(1) Set up a hasty defense. Adjust the perimeter as needed and make contact between subunits. Assign each a specific sector.

(2) Plan fire support. Plan FPFs on likely dismounted avenues of approach. Mass antiarmor weapons on the most dangerous mounted avenue of approach.

(3) Conduct reconnaissance. Identify enemy weak points. Locate enemy armored forces and the routes they must use. Locate enemy automatic weapons.

b. **Planning.**

(1) Organize all support personnel into maneuver elements--for example, company headquarters, mortars, and antitank. Organize the ranger force into a rupture force, reserve, support force, and a rear guard.

(2) Issue a fragmentary order. It should contain current local enemy situation and cover all pertinent areas of the five-paragraph field order.

(3) Select rally points.

c. **Execution.**

(1) Reduce the defense perimeter. This is done slowly until the unit is massed for the breakout attempt.

(2) Create a diversion. This is done either by fire from the rear guard force or by a limited attack in a direction other than that of the main breakout attack.

(3) Break through the defense with the rupture force. Hold the shoulders of the breakthrough.

(4) Move the reserve through to exploit the breakthrough.

(5) Move the support troops through with casualties. Have them collect casualties en route.

(6) Have the rear guard follow the support troops. It may have to fight a delaying action and use fire and movement to break contact with the enemy.

(7) Consolidate and reorganize following the breakout.

(8) Have the rupture force break contact with the enemy once the rear guard joins.

6-7. Linkup operations.

During operations behind enemy lines, ranger units may be required to link up with other friendly ground forces. Linkup may occur in airborne, air assault, amphibious, or shore-to-shore operations. It may occur during the breakout of an encircled force or during the convergence of separate forces. It may involve conventional or unconventional US forces, allied forces, or friendly guerrillas. Rangers may be required to link up with other rangers, partisans, or friendly ground forces that are conducting offensive operations.

a. **Planning.** Ranger commanders plan detailed linkup operations to avoid friendly forces firing on each other. Coordination and planning of communications, recognition signals, fires, command relationships and responsibilities, and control measures are essential. Coordination should be done before the start of the operation.

(1) Site. The linkup site should be easy to recognize, have cover and concealment, be located away from prominent terrain the enemy might use, be defensible, and provide multiple access and escape routes. An alternate site should be planned in case of compromise.

(2) **Communications and recognition signals.** Frequencies, call signs, codes, visual signals, and alternates for each of these should be planned before departure of friendly lines. Radios may be used to report the location of each unit to the site, and occupation and securing of the site. A system of mutual recognition must be devised to keep friendly patrols from firing on each other. This may include visual signals such as arm bands, panels, colored lights, or RSTA devices.

(3) **Fires.** The headquarters directing the linkup sets up fire control measures and priority of fires. The most commonly used control measure is the restrictive fire line (RFL), beyond which one force may not fire without coordination with another force. The RFL may be adjusted as two units draw closer together. Such an adjustment needs close coordination and should be planned before the operation. Successive phase lines between the two units can serve as on-order RFLS, provided that neither unit comes too close to the RFL before it is shifted.

(4) **Command relationships and responsibilities.** These are set up by the headquarters directing the linkup. Ideally, liaison personnel are exchanged before the operation. The key to success is liaison and guides.

b. **Execution.** Each member of the ranger force must understand the plan. Rehearsals are necessary in order to avoid confusion during the linkup. One uninformed ranger can start firing and cause enough confusion that friendly forces could be identified as enemy. Actions must be quick since units are open to enemy attack while they try to link up.

(1) **Moving units.** A linkup between moving units is hard. As the units move closer to each other, the chance of them engaging one another increases. Therefore, the linkup units must adjust their movements to each other and continuously coordinate on a designated secure radio net. If possible, one or both units should come to a short halt before linkup.

(2) **Stationary units.** Moving units must know the positions of the stationary units. They must keep stationary units advised of locations. Stationary units guide the moving units to the linkup point by radio. The stationary units must be ready to accept and position the moving units smoothly and quickly.

6-8. Passage of lines.

A passage of lines is an operation in which one unit moves through another unit that is stationary and disposed in a tactical formation on a FEBA. Ranger forces may conduct a passage of lines to get behind the enemy. Movement in forward unit areas must be controlled, coordinated, and kept to a minimum. This avoids conflict with friendly troops or the activation of RSTA devices. Ranger units treat the positions of forward units as danger areas. They must be assumed to be under enemy surveillance in all weather or visibility. The ranger force is open to attack during the passage. Personnel and units may be more concentrated than is desired. Fires of the stationary unit may be masked for a limited time. The passing ranger unit may not be in a good position to react to enemy contact. Detailed reconnaissance and coordination are crucial to ensure that the passage is conducted quickly and smoothly.

a. **Principles.**

(1) **Coordination.** The ranger force commander, or his representative, must coordinate departure and reentry of friendly front lines (FFL). This is done either directly with the units through which the rangers pass or through designated staff agencies. Items to be coordinated include fire support, passage lanes, control measures, guides, communications, intelligence, CSS, and actions on enemy contact.

(2) **Reconnaissance.** The ranger force commander should make a ground reconnaissance of the area through which the unit will pass. He should pick the initial rally point (IRP) at this time. The reconnaissance should include observing the area just forward of the friendly unit for routes, danger areas, and obstacles. During reconnaissance or coordination, rangers should wear the same uniform as the troops in the forward unit. This will avoid attracting the enemy's attention.

(3) **Security.** Maintain security to avoid contact with the enemy while departing or reentering. It is important that the ranger force make no enemy contact at this time. Control is hard if a fire fight starts just forward of an FFL.

b. **Departure of friendly front lines.**

(1) Designate an IRP.

(2) Maintain security. Use an appropriate movement formation when departing the FFL.

(3) Do not move within the friendly unit's area without a guide.

(4) Make final coordination with the friendly unit commander.

(5) Count the ranger force members as they depart.

(6) Make a security and listening halt.

c. **Departure of friendly units.**

(1) The ranger force arrives at the forward unit and is met by a guide. The guide leads the rangers to an assembly area chosen by the ranger commander during initial coordination with the forward unit commander.

(2) No one should move, either singly or as a unit, anywhere in the FFL without a guide.

(3) The ranger commander makes final coordination with the forward unit commander. He determines if any changes have taken place since initial coordination and learns of any recent enemy activity.

(4) Before leaving to coordinate, the ranger commander gives instructions (called a contingency plan) for what should be done while he is gone. These instructions state where he is going, who is going with him, how long he will be gone, what to do if he does not return, and actions to be taken if there is enemy contact.

(5) On the commander's return from final coordination, he may issue a fragmentary order (FRAGO) to cover any changes.

(6) Three common threats and techniques for countering them during departure from the FFL are:

(a) **Ambush and chance contact.**

- The ranger force commander must avoid enemy contact while departing.
- The ranger force goes to the friendly side of the departure point where a point team is dispatched to see that the area forward of the barriers is clear.
- The point element should check the area out to the first covered position or an area large enough to allow the unit room to maneuver if it is hit while departing.
- The distance varies with the size of the ranger force and the terrain.
- The point team notifies the ranger commander when the area is clear. The rest of the unit is brought forward.

(b) **Indirect fire.** If enemy indirect fire is falling, the ranger commander should not halt his unit after final coordination at the forward command post. Instead, he should move it quickly through the FFL. The point element should be far enough ahead of the main body to provide security from ambush or chance contact.

(c) **RSTA.** If the enemy has radar, sensors, or night vision devices, they may be offset by the following countermeasures:

- Use a well-hidden departure point, such as a reverse slope or a dense woods.

- Infiltrate the unit through the departure area and have the rangers rendezvous in a defilade.
- Pass through during rain, fog, snow, or darkness to help conceal the passage.
- Employ electronic countermeasures.

(7) Have a security and listening halt after the ranger force has moved out of sight and sound of the FFL. This is a short halt to see if enemy is in the area, and to accustom the unit to the sights and sounds of the battlefield. Halt in a position that covers the unit from chance friendly small-arms fire. If the security and listening halt must be in an exposed position, have each man lie flat on the ground.

(8) During movement, have frequent halts to see that the unit is not being followed and that no enemy is in the area.

d. **Reentry of friendly front lines.**

(1) Set up and occupy a reentry rally point (RRP).

(2) Maintain security at the RRP and at the reentry point.

(3) Use as few personnel as possible to locate the reentry point while the bulk of the unit remains in the RRP.

(4) Meet a guide at the reentry point. Normally, a forward-of FFL password is used since it may be overheard by enemy.

(5) Count the ranger force members to ensure no infiltrators follow the unit, especially during reduced visibility.

(6) Give the friendly unit commander a spot report, providing combat information of immediate tactical value.

e. **Reentry of friendly units.**

(1) The unit is moved into a rally point near the reentry point. This rally point should be on a prominent terrain feature where the ranger commander can pinpoint his location, with respect to the reentry point.

(2) The forward unit is alerted by radio that the ranger unit is ready to reenter. A code word is used for security and brevity. The code word must be acknowledged by the forward unit before the ranger force reconnoiters the reentry point. This shows that a guide has been sent to the reentry point and is waiting for the unit.

(3) If the ranger commander is certain of the reentry point location, he moves the entire unit directly to the reentry point. He has the point element coordinate the reentry.

(4) An effective method of locating the reentry point is through the use of a thermal imagery device. It detects the body beat of the reentry guide, even though he may be well camouflaged or his location masked by smoke. If ground surveillance radar is available, it can be used to vector the ranger unit to the reentry point.

(5) The reconnaissance party should never move parallel to friendly barriers or move around the wire. If the reentry point cannot be found by the initial reconnaissance or if the reconnaissance party finds only the barrier wire, the unit should notify higher headquarters and move to another rally point to wait until daylight. The unit should not stay in a rally point from which a radio transmission has been made.

(6) When the reentry point is found, the ranger commander can bring the unit forward. If the reentry point is easy to find, he can call by radio for the unit to come forward.

(7) The guide leads the unit through the barriers to the security position coordinated with the forward unit commander. The unit halts in the security position, remaining in movement formation. The ranger commander then gives a spot report to the forward unit commander. The ranger commander tells him anything of intelligence value or of immediate tactical use to the forward unit.

6-9. Reconnaissance platoon operations.

a. The reconnaissance platoon organic to the ranger regiment consists of a platoon leader, a senior reconnaissance sergeant, and three five-man reconnaissance teams that include a communications specialist in each team (see Appendix A). This platoon is responsible for short-duration reconnaissance of objectives for the ranger force commander. This reconnaissance is routinely in support of future strike or special light infantry operations. The teams can also be used by the ranger force commander to conduct deep target acquisition and designation, or to confirm or disprove information reported by other sources. Rarely will teams be used to attack enemy targets.

b. The reconnaissance teams of the ranger regiment do not replace (or perform the same reconnaissance tasks associated with) long-range reconnaissance patrol (LRRP) units or long-range surveillance units (LRSUs) that may be organic to the corps or division. These units are organized, trained, and equipped for long-term passive surveillance and reconnaissance in enemy territory. The ranger reconnaissance teams are trained and organized to support the immediate intelligence-gathering needs of the ranger force commander. They are not to act as a human intelligence (HUMINT) collection asset for other operational headquarters.

c. The reconnaissance platoon headquarters trains, equips, and controls the employment of the three reconnaissance teams. It operates the communications stations needed to receive reports from deployed teams. In the absence of the S2, it debriefs team members upon extraction or exfiltration. It can provide one or more teams to support each ranger battalion on independent operations. The reconnaissance teams can--

(1) Infiltrate the objective area by parachute (HALO, HAHO, or static lines), helicopter, fixed-wing aircraft, SCUBA, small boat, foot, or other means.

(2) Remain undetected in the vicinity of the objective area up to five days.

(3) Perform reconnaissance operations employing a full range of night observation devices, infrared detection devices, unattended sensors, and photographic equipment.

(4) Perform demolition target analysis.

(5) Operate small watercraft and inflatable boats.

(6) Emplace unattended ground sensors, omni-directional navigational beacons, hand-emplaced expendable jammers, and electronic target designation devices.

(7) Collect combat information to satisfy priority information requirements and mission-essential elements of information. Teams report that information by use of long-range, secure, burst-transmission communications equipment.

(8) Perform DZ selection, marking and reception duties.

(9) Report objective area weather conditions.

(10) Perform highly selective, limited attacks or ambushes when so tasked.

(11) Link up with the main body of the ranger force in the objective area, or escape and evade the enemy in order to return to friendly lines.

(12) Act as part of the AAT during airborne operations.

d. There are many limits of the reconnaissance teams.

(1) Mobility is normally restricted to foot movement in the area of operations.

(2) Use of radio and other active electronic or optical devices makes the teams open to enemy detection.

(3) All supplies and equipment needed by the team must be carried on the initial insertion, since resupply increases the probability of detection by the enemy.

(4) Conduct of insertion and extraction operations deep in enemy-held territory is difficult due to the need for secrecy, timeliness, security, and accuracy of location.

(5) Medical treatment of team casualties is limited to individual first aid. Casualty evacuation is difficult unless linkup with the main ranger force has been achieved.

e. The training of reconnaissance team members parallels that given to all members of ranger units with emphasis on infiltration, communications, and reconnaissance skills. Extensive training is also given in survival, evasion, resistance, and escape (SERE) and advanced first aid. Training emphasis is placed on actions at the reconnaissance site. The reconnaissance teams must conduct their mission and remain unseen. Every team member is trained to consider the need for information balanced by the likelihood of detection. Detection by the enemy compromises future operations.

6-10. Reconnaissance planning.

a. Detailed planning ensures that the reconnaissance mission is successful. It is also vital to the survival of the reconnaissance team. This planning is conducted by the ranger force planning group assisted by elements of the JTF staff. The supporting air or naval commander, navigator, and electronic warfare officer may aid the ranger force staff in the planning.

b. Upon receipt of the mission, the team selected begins an intense preparatory phase at the REMAB or the ISB. There the team receives its initial mission briefing. The items covered allow the team to complete its detailed planning. The reconnaissance platoon leader reviews the team plan. He ensures compatibility and integration into the ranger force commander's plan for employment of the main force. Staff specialists at ranger regiment and battalion--such as those concerned with intelligence, weather, or communications--provide help.

c. Planning for the use of reconnaissance teams is conducted within the framework of the ranger planning sequence for the entire mission. There are five planning considerations for the employment of reconnaissance teams:

(1) Operations security must be considered in all staff efforts to provide maximum protection for the reconnaissance teams. This includes intelligence, CE, logistics, administration, and maintenance. Operations security must be integrated throughout every ranger force operation. It consists of four main categories of security measures: signal security, physical security, information security, and deception. All are related and must be considered at the same time for each operation.

(2) Some missions may require rapid deployment to the operational area, thereby dictating the quickest method of insertion. However, others may depend on maintaining total secrecy, regardless of time.

(3) The transportation means selected for the delivery and recovery of the teams depends on the mission. This selection is also based on the abilities, limits, and availability of mission support platforms.

(4) The distance to and from the objective area must be considered, as well as the distance from the insertion point to the actual target. Reconnaissance teams can move great distances over rugged terrain. However, this movement does require time and limits the team's ability to carry bulky or sensitive equipment.

(5) Consideration must also be given to intelligence information available concerning the objective area which includes, but is not limited to--

(a) **Enemy situation.** The enemy threat--his abilities, disposition, security measures, detection, and air defense systems - affects the selection of the means for delivery or recovery.

(b) **Terrain.** Land formations must be considered in selecting the method of infiltration and exfiltration. Terrain affects the selection of altitudes, approach and exit routes, landing areas, drop zones, and beach landing sites. Air infiltration routes that provide terrain masking are desirable in static line parachute operations.

(c) **Weather.** Seasonal weather conditions affect team delivery and recovery. Factors to be considered are temperature, precipitation, visibility, clouds, and wind. If para-SCUBA techniques are used, high surface winds and their effect upon surf conditions or period of reduced visibility may preclude the use of parachutes, inflatable boats, or swimming as entry or recovery techniques. These same conditions favor land infiltration and exfiltration, The AWADS reduces the impact of visibility as a limiting factor for air infiltrations.

(d) **Astronomical conditions.** Periods of sunrise and sunset, moon phase, moonrise and moonset, and twilight must be considered.

d. A reconnaissance team mission must be specific and support the mission. Information concerning the target may not be gathered because the risk of detection by the enemy is so great. Detection may preclude the conduct of the planned operation. If more than one mission is assigned, priorities are set, All team missions must be coordinated to avoid other friendly forces operating near the target. Ranger forces conduct operations at such depth behind enemy lines that no other US or allied forces are normally present. There is a chance that a friendly indigenous guerrilla force may be operating near the objective, but ranger reconnaissance teams do not normally make contact with them.

e. The reconnaissance platoon leader joins with the intelligence and operations sections of the ranger force headquarters in the initial planning for reconnaissance team operations. The method of operation while moving, communication procedures, reporting, surveillance procedures, and actions on enemy contact should be included in the platoon's SOPS.

f. The ranger force intelligence officer normally prepares the reconnaissance plan IAW the guidance provided by the controlling headquarters. He is helped by the operations officer and the reconnaissance platoon leader. Selected team members and a member of the unit providing the insertion and extraction means are briefed on the mission early in the planning phase. They should participate in the detailed planning to follow. The essential details of a reconnaissance plan normally include--

(1) Areas to be observed and recommended areas or positions from which this can be done. These positions are determined in advance of the insertions of teams. They are based on a study of terrain, road and rail nets, enemy order of battle, delivery means, operations plans of the controlling headquarters, and the desires of the ranger force commander.

(2) Load plans and procedures. This includes both insertion and extraction. Alternate sites are chosen for possible use. Primary and alternate DZs are selected, if applicable.

(3) The flight plan. This includes insertion and extraction flight routes and alternate routes.

(4) The movement plan to and from a patrol base. This includes the times, routes,. and destinations when individuals or small groups are required to operate away from the patrol base.

(5) The fire support plan. This includes the location and times of planned air strikes in support of the main body, and all planned fires by the main assault force. The reconnaissance teams should know the frequencies and call signs for tactical air support.

(6) The timing and ground tactical plan for execution of the main force operation. This includes disposition of the main force, special mission teams from the ranger force, or any other special operations force elements.

(7) The communications plan. This includes frequencies, reporting schedule, emergency reporting procedures, and alternate communications plans. It also includes the ground-to-air

communications plan and the friendly recognition signal to be used during linkup with the main assault force or other friendly forces.

(8) Plans for use of technical specialists or special equipment.

(9) The plan for the treatment and evacuation of sick or wounded team members from the objective area.

(10) The E&E plan. This includes routes out of the objective area, passwords, and alternate extraction zones or methods far from the objective area.

(11) Mission-specific elements of information desired by the ranger force commander, and any special or on-order missions the team must conduct.

g. The team leader uses specific steps in planning, preparing, and executing missions. These procedures are comprehensive, yet adaptable to any situation. Three of the most important steps are:

(1) **Issuance of warning order.** This action gives the team maximum time to prepare for the mission. It should be issued as soon as the team leader has been notified of a mission and updated after he has made his tentative plan.

(2) **Issuance of OPORD.** The team leader conveys the information and instructions needed to accomplish the mission. He requires a briefback from each team member.

(3) **Rehearsals.**

h. The specific steps used when planning missions are as follows:

Step 1. Study the mission.

Step 2. Plan use of time.

Step 3. Issue initial warning order.

Step 4. Study terrain and situation.

Step 5. Make tentative plan.

Step 6. Organize.

Step 7. Update warning order.

Step 8. Coordinate (continuous throughout).

Step 9. Consult aerial photographs and map updates.

Step 10. Complete detailed plan. (To be combined with paragraphs 3, 4, and 5 of the operation order.)

Step 11. Issue operation order.

Step 12. Supervise (throughout), inspect, rehearse.

Step 13. Execute mission.

i. A workable E&E plan must be developed. Such a plan enhances the survival of the reconnaissance team members. It ensures that they are psychologically prepared to conduct the mission, by knowing there is a system and plan for returning them to friendly lines. The E&E plan must include a section describing the actions of the reconnaissance team members should their insertion craft be disabled en route to the objective area. It should also include what to do if the team is unable to link up with the main ranger force, or if the main ranger force has not been inserted into the objective area.

j. Rehearsals and briefbacks are best for finding flaws in procedures or planning. Thorough coordination of all procedures is essential. The exact type of aircraft or other means that will be used for insertion should be used for training. Rehearsals should be conducted under terrain, astronomical, hydrographical, and meteorological conditions close to those to be met on the operation. The team leader conducts a briefback with the platoon leader. This lets the platoon leader decide if the team is ready for the mission. It also allows him to make changes.

6-11. Execution of insertion.

a. Reconnaissance teams normally conduct blind insertions. Blind insertions are amphibious, heliborne, or parachute drops onto unmarked DZs without help from a DZ-marking party or a CCT. Blind insertions often depend on favorable light and weather conditions in the objective area. Favorable conditions for the reconnaissance team often means low-lying cloud cover, fog, and darkness. The teams are able to operate during reduced visibility by the extensive use of night observation devices. The team's extensive training and high level of land navigation skills allow it to quickly traverse rugged terrain and not be detected.

b. Team insertion into the objective area often needs a cover and deception plan. Team members can conduct HALO and HAHO operations. Teams can be inserted into areas not normally acceptable as a DZ for conventionally trained parachutists. By the use of HAHO techniques, reconnaissance teams can assemble in the air and glide to a landing at a location far from the drop point.

c. If reconnaissance teams need special or bulky equipment that cannot be carried on the individual team member, it can be airdropped, hidden near the DZ, and retrieved later. Such container drops are often used during low-level static line insertions. The high-altitude airdrop resupply system (HAARS) can be used to deliver such items at the same time as reconnaissance teams are inserted by HALO.

CHAPTER 7

Combat Support

Combat support is used to enhance the combat power of the maneuver companies. Combat support elements may be organic or nonorganic. They can conduct a variety of missions in providing indirect and direct fire support, security, intelligence, and maneuver and fire control.

7-1. Intelligence support.

a. The success or failure of a ranger mission often depends on the accuracy, detail, and timeliness of intelligence information. Both deliberate and quick-response missions need specific information that can be obtained only by a multidiscipline collection and analysis agency. An active interface with all levels of the intelligence system helps get the needed information quickly.

b. The theater or JTF commander is responsible for providing intelligence support. The 1st SOCOM also provides information and analyzed intelligence to the ranger force commander. The main source of processed intelligence is the Echelons Above Corps Intelligence Center (EACIC), similar to the division level military intelligence (MI) all-source production center (ASPC). The EACIC provides intelligence, security, and electronic warfare (EW) interfaces (see Figure 7-1). It is through the EACIC that ranger commanders can request support from national, joint, combined, or CONUS intelligence analysis systems.

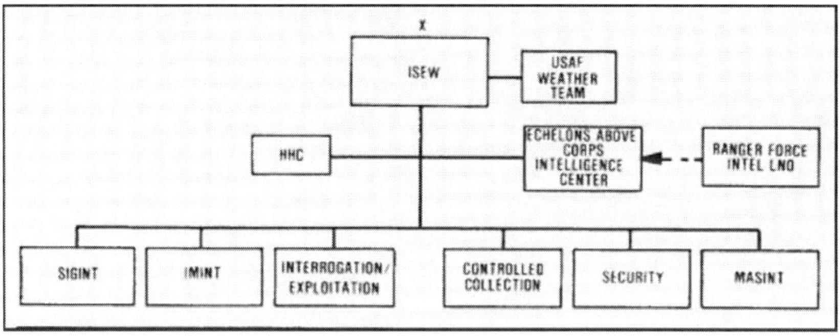

Figure 7-1. Intelligence at echelons above corps.

c. Intelligence requirements for ranger operations are unique, highly sensitive, and compartmented. The intelligence officer of the 1st SOCOM and the appropriate 1st SOCOM regional planning cell coordinate with theater intelligence agencies to provide needed information.

d. The ranger regimental staff and the ranger battalion staff have personnel in their intelligence sections to correlate and analyze information from many sources.

e. The reconnaissance platoon, organic to the ranger regiment, maintains surveillance of the objective area before the insertion of the main ranger force. It reports any last-minute developments that would affect the ground tactical plan. The platoon can also be tasked to determine any mission-specific items of essential information that the ranger force commander needs to complete his plan. This platoon is a highly specialized unit that is reserved for use by the ranger force commander. It is not a HUMINT collection agency for use by the theater or JTF commander. Reconnaissance teams normally link up with the ranger force in the objective area and are extracted with the main body. If this linkup cannot be made, they will exfiltrate enemy territory. They may use escape and evasion nets set up by other special operations forces or Department of Defense (DOD).

f. The ranger regiment normally places a liaison team at the headquarters of the command having OPCOM of a ranger force. This team provides command and control communications and liaison. It puts an intelligence liaison officer (LNO) at the EACIC or the appropriate ASPC (see Figure 7-2). This provides coordination to ensure that the intelligence needs of the ranger force are relayed to the intelligence processing center, and that the analysis is on the specific needs of the ranger force commander. This LNO coordinates with another located at the corps or EAC targeting center. Close coordination between the regiment, the intelligence analysis center, and the targeting center is essential to effective employment.

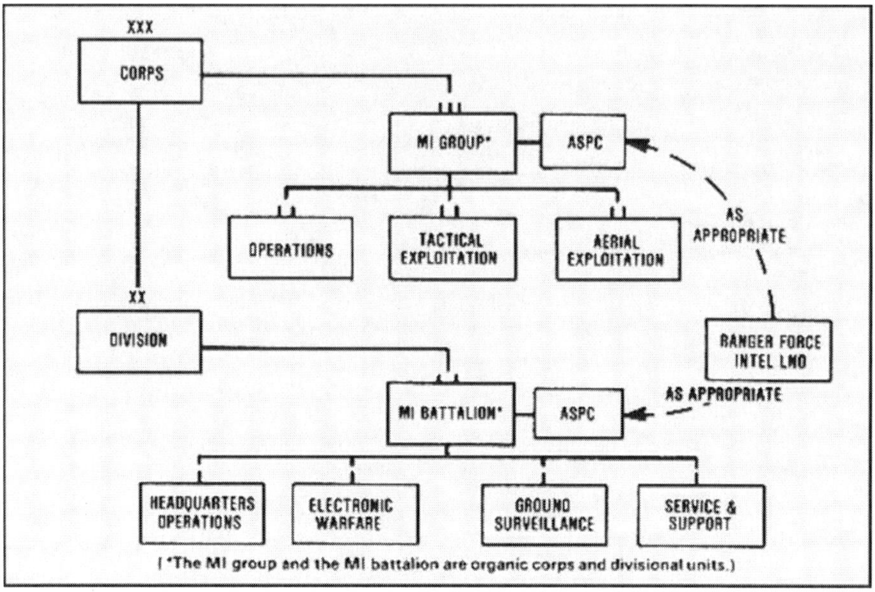

Figure 7-2. Intelligence source at corps and division.

7-2. Electronic warfare support.

a. Depending on the nature of the target and enemy abilities, a ranger mission may need support from EW units. This is mainly true during insertion. Electronic warfare operations disrupt or destroy the enemy's command and control of his forces and weapons systems, and retain friendly use of the electromagnetic spectrum. Electronic warfare is also used to support deception operations. They mislead the enemy by manipulation, distortion, or falsification of indicators, and persuade him to react in a manner prejudicial to his own interests.

b. Active jamming and chaff dispersal can prevent enemy early-warning radar from detecting the airborne force or determining its route. False transmissions can aid deception plans or feints used to help the ranger force's entry behind enemy lines. Selective jamming and imitative deception can disrupt the enemy command and control nets. This prevents the enemy from reacting to the presence of the ranger force in time to prevent mission accomplishment, or slows and interrupts his deployment of reaction forces.

c. Ranger elements, or other military forces, may conduct closely coordinated direct attacks against enemy EW sites during a ranger force's insertion or extraction. Not only does this reduce the enemy's ability to direct EW against the ranger force command, control, and communication system, but it misleads and confuses the enemy as to the true ranger objective.

d. The JTF or theater commander is responsible for planning the EW program to support a ranger mission. Planning and coordinating elements from the 1st SOCOM may also help. The EW program must be closely coordinated so as not to disrupt friendly air-to-ground or ground-to-ground communications, or to reveal the ranger force intentions.

e. The ranger regimental communications-electronics (CE) officer and those of the ranger battalions ensure that the required coordination concerning jammed frequencies, codes, authentication tables, and so on, is completed. The ranger regimental and battalion operations personnel ensure that the efforts of all EW units are combined to support the ground tactical plan.

7-3. Communications support.

a. Ranger operations are supported by secure long-range, lightweight, real-time, high-frequency (HF), and satellite communications. Effective long-range communications provide command and control links between deployed ranger units and the controlling headquarters. The JTF, ARFOR, or theater commander is responsible for communications between the controlling headquarters and the ranger regiment. The special operations force commander may also provide secure communications terminals to the ranger regiment or a deployed battalion.

b. Secure AM, FM, and SATCOM radios are the primary means of communication within the ranger regiment. Within the ranger battalion, AM and FM radios provide communications to company, platoon, and squad level.

c. During certain missions, specially trained and equipped quickreaction elements (QRE) deploy with the ranger force to provide secure communication links to the special operations force commander. These JCSE elements operate on either SATCOM or tactical satellite terminal (TACSAT) channels or through an airborne communications relay platform (see Figure 7-3). Depending on the mission, an airborne battlefield command and control center (ABCCC) aircraft or a joint airborne communications center/command post (JACC/CP) may be used. Their communications systems can operate at all levels of the national chain of command to permit a quick response to the tasking authority.

Figure 7-3. Airborne communications relay platform.

d. When only one ranger battalion is employed and the regimental headquarters is not the controlling headquarters, the ranger regiment provides a liaison cell to the controlling higher headquarters. This liaison cell includes a communications element from the regimental communications platoon. It can provide secure SATCOM, teletype, and facsimile transmission support to the ranger force in the objective area.

e. If two or more ranger battalions are employed, the ranger regimental headquarters deploys and acts as the command and control headquarters. The regimental communications platoon would then provide another communications link to the special operations force commander.

f. The ranger regimental CE officer ensures that the necessary communications links are set up and coordinated. The many communication means and channels available provide for effective control of a deployed ranger force. However, communication means must be closely coordinated at all levels of command to control the complex operations of a ranger mission. Planning and coordination with supporting aviation, transport, fire support, medical, and logistical elements before an operation are vital to efficient communications. The regimental CE officer must consider the communication systems linking the ranger force and other services. Air-to-ground and ship-to-shore communications are vital and must be set up early in ranger operations.

7-4. Fire support.

a. **Planning.**

(1) The ranger regiment has limited organic fire support assets. The six 60-mm mortars in each ranger battalion (two in each rifle company) provide the only organic indirect fire

available to the battalion commander. Each rifle company also has three 90-mm recoilless rifles and three Dragon missile systems that provide a small measure of organic direct fire support.

(2) The missions assigned to the ranger regiment and its subelements normally require fire support from organizations outside of the regiment. To plan, coordinate, and control these fires, the ranger regiment is organized with personnel dedicated to these tasks.

(3) The ranger regimental headquarters has a fire support element consisting of one fire support officer (FSO), two noncommissioned officers, and two fire support specialists. They prepare the fire support annex to the regiment's operation order. They ensure that all available fire support is planned to best support the regimental commander's ground tactical plan. They act as the primary coordinator for all external fire support, such as artillery, armed helicopter, close air support, naval gunfire, and short-range air defense. They also coordinate with each ranger battalion FSO to ensure unity of effort. The fire support element coordinates the efforts of air force, naval, and marine liaison officers when they are attached.

(4) The regimental commander may assign part of the fire support element to a committed ranger battalion as an attachment to help in the integration and planning of fires. The ranger battalions request through the fire support element that more or different fire support means be integrated into the ground tactical plan.

(5) The ranger battalion headquarters has a fire support team headquarters consisting of the battalion's FSO, two noncommissioned officers, and two fire support specialists. The FSO is the prime fire support coordinator for the ranger battalion commander. He maintains coordination with the regimental FSO and any attached fire support coordination elements, such as the USAF tactical air control party or naval shore fire control party. The FSO performs the following:

(a) Advises the ranger battalion commander on all fire support matters.

(b) Recommends allocation of fire support.

(c) Prepares fire support plans.

(d) Assigns target numbers.

(e) Processes target lists from the ranger company fire support teams to eliminate duplication.

(f) Monitors and functions as net control on the ranger battalion fire support coordination (FSCOORD) net.

(g) Reports changes in the status of fire support units to the ranger battalion commander, staff, and FISTS.

(6) When the ranger battalion is operating a single TOC, the FSO and his element are normally located there. When the ranger battalion is operating two TOCS, one TOC is normally manned by the FSO, the fire support sergeant, and a fire support specialist. The

senior fire support sergeant and the other fire support specialist would then monitor the FSCOORD net at the other TOC. The FSO ensures that any attached fire support elements also divide when needed to operate in a two-TOC configuration.

(7) The ranger battalion fire support team headquarters has three fire support teams assigned to it. These are normally allocated one to each of the ranger rifle companies. The company FSO and his team--

(a) Locate targets and request and adjust surface-to-surface fire support (mortar, field artillery, and naval gunfire).

(b) Plan fires to support the company ground tactical plan and prepare target lists.

(c) Coordinate fire support requests through the FSO for surface-to-surface and air-to-surface fires.

(d) Report battlefield information.

(e) Control air-to-surface fires of helicopters and prepare to assume control of fires from fixed-wing aircraft.

(f) Coordinate airspace use within the ranger rifle company operational area.

(g) Advise the ranger company commander on all fire support capabilities, limitations, and methods of employment.

(h) Inform all fire support units of target priority changes.

(i) Designate targets for laser-guided munitions.

(j) Assume operational control of, and administrative responsibility for, all augmentation fire support personnel (TACP, shore fire control party, and attack helicopter liaison teams).

(8) The company FSO is normally with the ranger rifle company commander. The FOs and their RATELOs are normally with the ranger rifle platoon leaders.

b. **Coordination.**

(1) Coordination of fires is a continuous process. It involves implementing the fire support plans of the ranger force commander and reacting to changing tactical situations. The success of a mission often hinges on how well fires are coordinated during the operation.

(2) The fire support planners within the many echelons of the ranger force become fire support coordinators during the actual execution of the operation. They use standard fire control coordination measures.

c. **Organic fire support.**

(1) **Indirect fires.**

(a) The 60-mortars constitute the only organic indirect fire support available within the ranger regiment. The mortar squads in each ranger rifle company normally operate as a section under the control of the mortar section sergeant. However, they can operate alone for short periods.

(b) The 60-mm mortar is used to destroy enemy equipment and light vehicles and to attack enemy personnel. Its main advantage is the ease with which it can be brought into operation against the enemy. This ease of employment allows the ranger force commander to concentrate combat power at the critical time and place. This helps to surprise and destroy the enemy.

(c) Mortar fires are employed by the ranger rifle company commander and coordinated by the company FSO. They are normally requested and adjusted by the FIST observers. The mortars are normally emplaced in a central position, and their fires are controlled by the fire direction center (FDC) using the indirect-fire method. The mortars may also use the direct-alignment or the direct-lay method.

(2) **Direct fires.**

(a) Each ranger battalion is authorized 90-mm recoilless rifles and Dragon missile launchers. These weapons constitute the heaviest organic direct-fire ability in the regiment. The antitank section in each ranger rifle company may operate as a complete section, but normally individual weapons are assigned throughout the company sector wherever their fires can best support the mission. These weapons can destroy most enemy armored vehicles and can penetrate the flank and rear armor of all enemy main battle tanks. The 90-mm recoilless rifle has a multiple projectile round that is effective against exposed enemy personnel. The HE round is highly effective against buildings and bunkers.

(b) The main disadvantage of direct-fire antiarmor weapons is their weight. These weapons, and their ammunition, are man-portable but place a strain on the unit when moving across rugged terrain.

(c) The ranger battalion is trained in the use of light antiarmor weapons, such as the US M72A2 LAW and the Warsaw Pact rocket-propelled grenade (RPG). The following provide protection for the ranger force against attack by enemy armored forces:

- The employment of light antiarmor weapons at all organizational levels.
- The placement of the 90-mm recoilless rifles or guided-missile systems on the most likely avenues of enemy-armor approach.
- An aggressive, offensively oriented employment.
- The ability to move into and across terrain impassable to armored vehicles.

d. **Field artillery fire support.**

(1) Many ranger unit operations are out of the range of supporting field artillery fires. Whenever such fires are available, they are planned for and integrated into the ranger force ground tactical plan.

(2) Coordination and exchange of call signs, frequencies, and target lists occur before insertion of the ranger force. Unless required by the tactical situation, ranger units do not normally receive augmentation forward observers from supporting field artillery units. The degree of mental and physical training necessary to execute certain ranger techniques requires more preparation than field artillery units are able to provide. The FIST from the ranger battalion requests and adjusts field artillery fires for the ranger force.

(3) Field artillery fires can be used to support the ranger force even if the objective area is out of range. Field artillery cannon fire and multiple rocket launch system (MRLS) fire can be used to suppress enemy air defenses. This helps the ranger force as it crosses the FEBA during airborne or air assault operations.

(4) Field artillery fires can be used to support the exflltration of ranger elements as they approach friendly lines. These fires can contribute to the deception plan and add combat power to feints used to support ranger operations.

(5) When in range of the objective area, field artillery units can be used to emplace field artillery delivered FASCAM to enhance the security of the ranger force.

(6) Copperhead rounds fired by 155-mm field artillery units can be terminally guided by the ranger FIST forward observers (FOs). They can attack hardened point targets or enemy armored vehicles by using a man-portable laser target designator (see Figure 7-4).

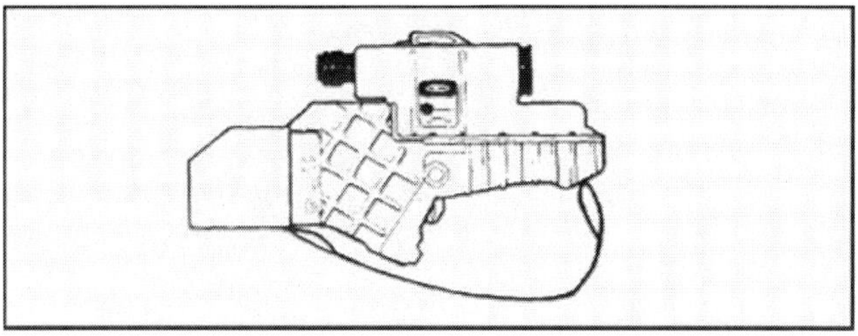

Figure 7-4. AN/PAQ-1 laser target designator.

e. **Aerial fires.**

(1) Aerial fire support is usually the prime means supporting the ranger force due to the distance behind enemy lines at which most ranger operations take place. Aerial fire support can be provided by either fixed-wing aircraft or helicopters.

(a) Fixed-wing aerial fire support may come from USAF, USN, or USMC units. The type of unit providing support, the aircraft, and the mix of ordnance carried all affect the fire support planning and coordination process. Some aircraft have a night and all-weather strike ability enabling them to support the ranger force during any level of visibility. Operations during poor weather that limits visibility to less than 3 nautical miles are still somewhat restricted. The ranger fire support coordinators must ensure that the correct aircraft are requested and employed effectively on the enemy. The

TACP directs and adjusts aerial fires in the objective area.

(b) The ranger force can use ground laser target designators to pinpoint targets for air strikes, as well as electronic navigation aids to permit nonvisual air strikes (beacon bombing). The ranger rifle company FIST or the TACP can control a laser-designated standoff air strike (see Figure 7-5).

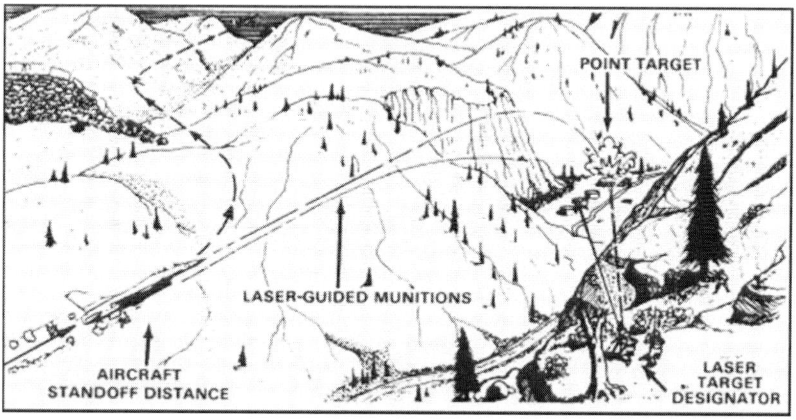

Figure 7-5. Standoff air strike.

(c) If the enemy ADA ability is not great or it can be degraded to a low level, the ranger force uses specially equipped and armed AC-130 aircraft (see Figure 7-6) for fire support. These aircraft provide an invaluable combination of firepower, night observation and illumination, communications, and long loiter time. A well-planned and executed suppression of enemy air defense (SEAD) program, coupled with ECM directed against enemy ADA units, normally permits the use of AC-130 aircraft.

Figure 7-6. AC-130 aircraft.

(2) The attack helicopter, armed with a mix of antitank guided missiles (ATGM), 2.75-inch rockets, 20-mm cannon, and 40mm grenade launchers, is accurate and responsive aerial fire support. The ranger force commander plans to use all sources of aerial fire to help accomplish his mission. However, the nature of ranger operations may preclude the extensive use of armed helicopters due to their limitations. Some attack helicopters are limited in range and lack an all-weather ability. They are restricted during operations at night by a lack of sophisticated night vision devices. They may lack the speed to go with the ranger force on some heliborne insertions. Attack helicopters may be used to escort and assist the ranger force as it crosses the FEBA. They may also be used to conduct feints and demonstrations to cover the insertion of the ranger force. As the MV-22 Osprey and the AH-64 Apache enter service, more use will be made of these aircraft to support ranger operations beyond the FERA, (See Figures 7-7 and 7-8).

Figure 7-7. MV-22 Osprey.

Figure 7-8. AH-64 Apache attack helicopter armed with Hellfire missiles.

(a) If attack helicopters are used to support a ranger operation, planned indirect fires are normally delivered along entry and exit corridors. Attack helicopters approach and depart the objective area using nap-of-the-earth flight profiles.

(b) Fires from armed helicopters are normally requested and controlled by the company FSO or one of his FOs, operating on a special ground-to-air net. The laser target designator may be used to precisely identify targets for the AH-64 Apache. Friendly unit locations may be marked by smoke, panels, lights, mirrors, or infrared sources.

f. **Naval gunfire.**

(1) During amphibious assaults or operations near a coastline, the ranger force may receive indirect fire support from naval gunfire. Normally, a ranger battalion would be supported by either a destroyer or a cruiser in a direct support role. However, the type and importance of the mission, the type of targets, the ships available, the hydrographic conditions, and the enemy capability determine how many and which type of ships are provided to support the ranger force.

(2) Naval gunfire is characterized by large volumes of highly destructive, flat-trajectory fire. Planned strikes in support of the ranger force may also include surface-to-surface missile fire. Some naval guns can fire a laser-guided projectile much like the Copperhead. This projectile, called a Deadeye round, is terminally guided the same as the Copperhead. Deadeyes can be guided either by an attached spotting party from the ANGLICO or by ranger LTD teams.

(3) When ranger units are being supported by naval gunfire, a shore fire control party is attached to the ranger force, This party normally consists of a liaison team and several spotter teams (forward observers). The liaison team is integrated into the operations of the fire support element (FSE) at the ranger battalion TOC. The spotter teams are attached to the ranger rifle companies. The shore fire control party LNO is the ship's representative to the ranger force commander, through the FSO. The spotter teams request and adjust fires from surface vessels. They can also request and control air strikes by carrier-based aircraft.

(4) The LNO and the spotter teams operate in the ground spot net, communicating with the ship by high frequency (HF) radio to request and adjust naval gunfire. The spotter teams communicate with the LNO using very high frequency (VHF) radios. The LNO also can communicate with aircraft using ultrahigh frequency (UHF) radios.

(5) Coordination and control measures that apply to naval gunfire are the same as for field artillery, with the addition of two terms, which are:

(a) Fire support area (FSA)--a sea area within which a ship may position or cruise while firing in support. It is labeled as FSA and numbered by a Roman numeral - for example, FSA VII.

(b) Fire support station (FSS)--a specified position at sea from which a ship must fire. This is very restrictive positioning guidance. It is labeled as FSS with a Roman numeral--for example, FSS VII.

g. **Air defense artillery.**

(1) Army air defense artillery units are rarely used to directly support the ranger force since

ranger operations normally take place deep behind enemy lines. However, it is possible that during the insertion or extraction phase of an operation, ADA units may support the ranger force as it crosses the FEBA.

(2) The ranger regiment's fire coordination element and the FSO in the ranger battalion are mainly concerned with adequate coordination of ADA fires. This prevents friendly units from being engaged. It allows the integration of ADA fires into the scheme of maneuver. The fire support planners ensure that enough coordination has been made so that ADA units are aware of the ranger force and know enough about its flight paths, routes, and altitudes to prevent engagement by friendly units. When conducting this coordination, care must be taken to pass only needed information to the ADA unit. To aid coordination, an ADA officer may be attached to the ranger force.

(3) The plan for a heliborne or airborne insertion and extraction of the ranger force must include the correct identification-friend-or-foe (IFF) codes, the flight profiles, the times of FEBA crossings, and the estimated number of aircraft making the crossing. There must also be a plan for crossing by straggler aircraft and battle-damaged aircraft that cannot reply to IFF interrogation. If possible, the routes selected should avoid concentrations of friendly ADA units.

(4) During an aerial crossing of the FEBA by a ranger force, the regimental headquarters may assign a liaison officer to the headquarters controlling the ADA fires in the crossing area. The LNO is then responsible for close and continual coordination. This ensures that the crossing is made without friendly ADA units engaging the ranger force while it is exiting or entering friendly lines.

(5) During ranger operations, the primary source of air defense fires is shoulder-fired antiaircraft missiles carried by specially trained individual rangers. These weapons provide the ranger force commander a limited defense against attack by enemy aircraft. The decision to take these weapons into the objective area depends on the factors of METT-T. Ranger operations should be completed before the enemy can react with ground forces. However, enemy aircraft may be able to react in time to threaten the ranger force. These lightweight missiles offer a highly effective, short-range counter to enemy air attack.

(6) The ranger force commander must define the weapons control status and rules of engagement for all short-range air defense (SHORAD) teams under his control. Priorities of target engagement must also be determined to prevent multiple engagements of a single enemy aircraft. The ranger battalion FSO and the attached ADA officer are responsible for the coordination of ADA fires within the objective area. Standard ADA weapons-control status and rules of engagement may be modified by the ranger force command.

7-5. Aviation support.

a. **Planning.**

(1) The ranger regiment may need extensive aviation support on some missions. The ranger battalion is organized and trained to move quickly to the objective area by air and to be resupplied by airdrop for short periods.

(2) Often the decision to execute a mission depends on the amount and type of available insertion and extraction aviation support. Normally, the theater commander or JTF commander allocates the type and amount of aviation assets to support a ranger operation. The theater or JTF commander's staff is responsible for the planning and coordination concerning aviation support. The ranger force commander must play an active part to ensure mission success.

(3) Due to the extensive training requirements and specialized operational techniques needed to support special operations, special air operations units in the Army and the USAF have been equipped and trained for the mission. These forces train with the ranger regiment and jointly develop tactics, techniques, and procedures of employment. These special units operate diverse aircraft that have unique capabilities. These help the ranger force commander to effectively employ his units. Command of these special units is maintained at a very high level in the chain of command. They are assigned to support a specific operation. Then they return to their parent unit's control.

b. **Coordination.** In addition to the staff of the theater commander or JTF commander, the coordination of aviation support occurs at all levels in the ranger force. These operations often are conducted in phases. A separate air mission commander is then designated. The AMC is responsible for the completion of each phase of the aviation mission, as well as the close and continuous coordination with the ranger force, commander. The ranger regimental air operations officer and the movements control officer are responsible for the planning, integrating, and controlling of aviation support.

c. **US Air Force.**

(1) The USAF provides special aviation support to the ranger regiment and other elements of the special operations force. The USAF has elements that are equipped with MC-130 Combat Talon aircraft (see Figure 7-9) and AC-130 Spectre gunships, as well as HH-53 Pave Low, UH-60, and UH-1 helicopters. These aircraft are specially equipped to conduct long-range infiltration, exfiltration, attack, and search-and rescue operations. The USAF also provides C-141B and C-130 aircraft for strategic and tactical airlift of the ranger force. There are specially trained USAF crews that can operate using AWADS and SKE. This gives them an excellent means of inserting the ranger force during reduced visibility.

Figure 7-9. MC-130 Combat Talon aircraft.

(2) The AFFOR commander provides the airlift control element (ALCE) support at the departure airfield. The ranger air operations officer ensures that coordination of aircraft load limits, air movement times, specific load plans, and manifests is conducted in an effective and timely manner. The air operations officer and the movements control officer ensure that a departure airfield control officer (DACO) from the deploying unit is designated to provide face-to-face coordination at the departure airfield.

(3) The USAF normally provides an airborne battle command and control craft. This is a specially equipped C-130-type aircraft that provides communications and data system down-links to ground stations.

(4) The terminal guidance needed for assault landings and air traffic control in the objective area is provided by a specially trained and equipped CCT. This element has habitual training and operations relationships with the ranger units. The USAF is also responsible for the specially trained and equipped weather teams that can be sent into the objective area with the CCT.

(5) The USAF Tactical Air Command (TAC) includes fighter, interceptor, ground attack, electronic countermeasure, early warning, reconnaissance, and airborne command and control squadrons, any of which may be used to support the ranger force. The Tactical Air Command provides the results of aerial reconnaissance to the AFFOR, who then processes it for use by the theater or JTF commander's staff. This reconnaissance may be by low-level photo, imaging infrared, side-looking airborne radar (SLAR), low-light-level television, or overhead photography.

(6) The Military Airlift Command (MAC) equips and trains special long-range SAR aircraft and crews. These aircraft, mainly helicopters, are equipped with sophisticated navigation, observation, and communications equipment, as well as extended-range fuel tanks and in-flight refueling capabilities (see Figure 7-10). The ranger regiment uses these aircraft for certain missions.

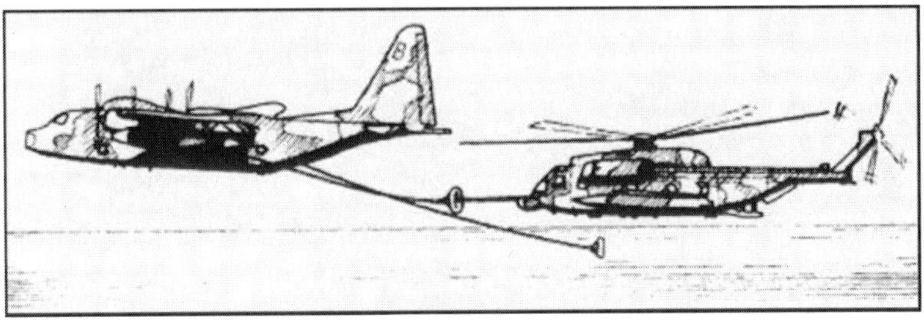

Figure 7-10. In-flight refueling.

d. **US Navy and US Marine Corps.** Both the USN and the USMC have units equipped and trained to support ground forces. The type of aviation support normally provided to the ranger regiment is close air support against targets in the objective area, or as suppression missions against enemy air defense installations. Long-range helicopters from the USMC may be used to support an amphibious operation or to insert special teams during ranger operations. The USN also has a search-and-rescue ability that may be used to support the ranger regiment.

e. **Special operations aviation units.**

(1) The US Army has specially equipped and trained SOA units tasked to support the ranger regiment, as well as other elements of the special operations force. These SOA units have a habitual training association with the ranger regiment and use dedicated aviation assets to -

(a) Insert, extract, and resupply the special operations force.

(b) Conduct armed escort, reconnaissance, surveillance, and electronic warfare in support of the special operations force missions.

(c) Provide airborne command, control, and communications for special operations force elements.

(d) Provide general support aviation during peacetime and contingency operations.

(2) Penetration of enemy airspace is normally done during limited visibility at very low altitudes. Penetrations are planned to maximize the use of existing terrain and exploit blind spots in enemy air defenses. On-board navigational aids are crucial to mission success. There normally is not any terminal guidance to the selected LZ or DZ. Deception operations of a tactical or operational nature help SOA units penetrate to the objective and reenter friendly lines. Advanced helicopters and short-takeoff-and-landing aircraft provide the help needed to do these missions.

(3) To survive enemy air defense systems and prevent mission compromise, SOA units avoid detection. Detected aircraft may reveal the location of supported special operations force ground forces or may be destroyed by enemy ADA fires. The SOA normally avoids detection by operating at night without illumination. This condition requires flight using sophisticated night vision devices. The SOA also operates in other conditions of reduced

visibility, such as clouds, fog, rain, snow, smoke, and dust storms.

(4) Before an operation, the ranger force commander may need to place reconnaissance teams near the objective. Army SOA assets can be used to transport ranger reconnaissance teams.

(5) Elements of SOA provide the ranger regiment with short and intermediate range penetration, resupply, and extraction capabilities. Long-range aircraft must be provided from USAF special operations assets.

(6) The ranger regimental or battalion commander is normally provided a liaison aircraft while the ranger force is in the REMAB. This ensures that he can travel quickly to the dispersed headquarters of the agencies providing support for the mission. This aircraft and crew are provided by the theater or JTF commander from EAC general support aviation assets. It may be a helicopter or fixed-wing aircraft, depending on the location of the REMAB to a suitable landing field.

(7) Army helicopters and fixed-wing aircraft may be used to provide SEMA aerial reconnaissance; radar, infrared and photographic imagery; and electronic intelligence support to the ranger force commander. This support is normally provided by the aerial exploitation battalion of the theater Army military intelligence group.

7-6. Engineer support.

a. The ranger regimental and battalion operations officers plan and request engineer support for ranger operations.

b. The most common engineer-type tasks needed during a ranger operation are the destruction of enemy facilities, equipment, and material by use of conventional explosives; and the interdiction of major lines of communication. To reduce the need for external support, the ranger regiment trains selected individuals in advanced demolition techniques. All members of the ranger regiment are well trained in the basic employment of conventional demolitions. Therefore, there is no need to attach combat engineers to conduct normal demolition missions. If specialized skill or knowledge is needed for a specific mission, qualified personnel can be attached.

c. During the occupation of the REMAB and the rehearsal for an operation, the ranger regiment needs construction engineer support to build simulated target areas. The simulated target areas should be camouflaged to prevent the enemy from recognizing the target. The construction engineers augment the ranger unit's self-help construction efforts. Limited-facilities construction and improvement may also be needed at the REMAB to provide security and isolation. The theater commander or JTF commander provides the construction engineer support.

d. Certain missions, such as those of interdiction, may need engineer special weapons teams. The teams are provided by the theater commander or JTF commander, or they may come from other special operations force assets. They are attached to the ranger force for a specific mission and released at the completion of the extraction and debriefing phase.

e. Some operations may need the ranger force to use captured enemy heavy construction equipment. Selected members of the ranger regiment are trained to operate such equipment.

However, construction engineer personnel may be attached to the ranger force for this mission.

CHAPTER 8

Combat Service Support

Combat service support consists of the logistical and administrative effort to maintain the battalion's ability to fight. The commander forecasts his needs for CSS. He employs CSS units to ensure accomplishment of the mission.

8-1. Ranger support element.

a. The ranger support element is a non-TOE organization formed from elements of CSS units collocated with CONUS-based ranger units augmented by selected personnel from the host installation. The mission of the RSE is to provide all the CSS needed to outload and deploy a ranger unit from CONUS. The personnel designated as part of the RSE have the same alert notification, reaction, and security requirements as members of the ranger unit they support. The RSE is normally built around a command and control group provided by a major CSS unit on the ranger unit's host installation. The remaining CSS elements and teams are then designated. An RSE consists of--

(1) An emergency operations center liaison group that provides timely coordination between the ranger unit and the functional elements of the RSE.

(2) A command and control group that coordinates and directs the actions of the functional elements.

(3) A supply section responsible for the timely issue of all Class II, III, IV, and VII items. This section maintains a stockage of contingency items marked for use by the resident ranger unit. It also issues items out of common stocks.

(4) A transportation section moves equipment and personnel from the unit area to the departure airfield and picks up equipment from local storage sites. It may also operate material-handling equipment and trailers to support loading of aircraft.

(5) An ammunition section responsible for the requisition, storage, issue, loading, and security of the ranger unit's basic load(s) of conventional munitions.

(6) A mess section provides all Class I support to the ranger unit. This includes operating a dining facility, providing food to remote sites, and drawing and issuing operational rations.

(7) A load team loads and unloads all equipment and supplies for the ranger unit. This team prepares pallets for loading, documents hazardous or dangerous cargo, and loads equipment pallets on USAF aircraft. It can also provide limited security for loads.

(8) A communications and electronics section provides secure communications among the RSE subelements, ranger unit, EOC, and DACG.

(9) An engineer platoon that is responsible for the construction of targets and rehearsal areas. This element of the RSE may or may not be activated, depending on the mission given the ranger unit. If the ranger unit deploys from CONUS to the objective area, this platoon has a key role to play in the premission preparation.

(10) A maintenance team whose primary mission is to ensure that the vehicles used by the RSE remain operational. It also provides predeployment maintenance and Class IX support for ranger unit vehicles and equipment.

(11) A medical team that provides unit-level medical care while the organic ranger medical section completes its outloading preparations. This team also provides any Class VIII supplies that the ranger unit may need.

(12) A security team responsible for security at the EOC, ranger unit location, and departure airfield. This team enforces stringent physical security precautions and maintains a constant counterintelligence (CI) surveillance of the outloading procedure. The security team provides armed guards for the protection of ammunition convoys, equipment movement, and vehicles.

(13) A rigger detachment that maintains, packs, rotates, and issues the personnel and equipment parachutes required by the ranger unit. It maintains enough stock of personnel parachutes to conduct a battalion-level parachute assault. The riggers lend technical skill to the rigging and aerial delivery of bulky, unusual, or delicate equipment.

b. The primary mission of the RSE is to provide the support necessary to quickly load and deploy the ranger units. The theater commander provides CSS to the ranger force once it has arrived at the OCONUS REMAB or ISB. The transition of support from the RSE to the CSS assets of the theater commander may be delayed for certain subelements in order to ensure a smooth deployment and enhance mission accomplishment. These subelements are released when the in-theater CSS assets can provide the support. Retained subelements of the RSE may be released from support requirements all at once, released in increments, directed to continue to provide support until mission completion, or augmented by in-theater CSS assets. The maintenance, medical, security, and rigger elements are those most likely to continue providing support OCONUS. This is likely if deployment is to an austere ISB pending immediate employment.

c. Once a ranger unit has been deployed OCONUS and has closed to a REMAB or ISB, the theater commander's staff coordinates CSS. If 1st SOCOM elements have deployed to the theater or operational area, the special operations support battalion (SOSB) of the 1st SOCOM coordinates CSS. Ranger logisticians at regiment and battalion level are mainly planners with limited logistics operations ability. Ranger logistical and personnel officers coordinate with CSS elements who provide such services.

d. The Theater Army Materiel Management Center (TAMMC) is the point of contact for information concerning in-theater logistical assets. In a developed theater of operations, a series of theater Army support groups will have been set up. One of these groups is assigned the mission of supporting the ranger force. The 1st SOCOM may also deploy certain assets available to provide CSS to the ranger force. The many CSS sources, the rapidly changing nature of SOF operations, and the unique CSS requirements of ranger units combine to make providing logistical and personnel support a challenge. The ranger unit's logistics officer must be included in the early phases of planning and deployment.

8-2. Logistics and personnel.

a. The echelons-above-corps CSS structure provides support to the ranger regiment as directed by the theater Army (TA) commander. In a low-intensity contingency involving a corps or less, there may be no echelons-above-corps logistics and personnel support elements. The corps support command (COSCOM) involved may be able to meet all needs from its own sources. However, in a combined operation in a high-intensity environment, the theater commander assembles a large CSS organization to provide support to the TA. This CSS base normally provides for all the ranger unit's requirements. During contingency operations in an area where there is no TA, the JTF commander provides the CSS elements. Until the buildup of these elements is complete, the ranger regiment must be prepared to operate from austere facilities with only the supplies it brings from CONUS.

b. The ranger regiment is normally assigned to the theater command for CSS. The TA commander, as directed by the theater commander, provides CSS to the ranger regiment and sets up support priorities. The EAC CSS organizations are under the command of the TA commander and consist of subordinate area and functional commands. The TA plans, coordinates, and executes its CSS functions through area-oriented support commands and specialized subordinate organizations. During the early stages of conflict, the TA may be organized as shown in Figure 8-1.

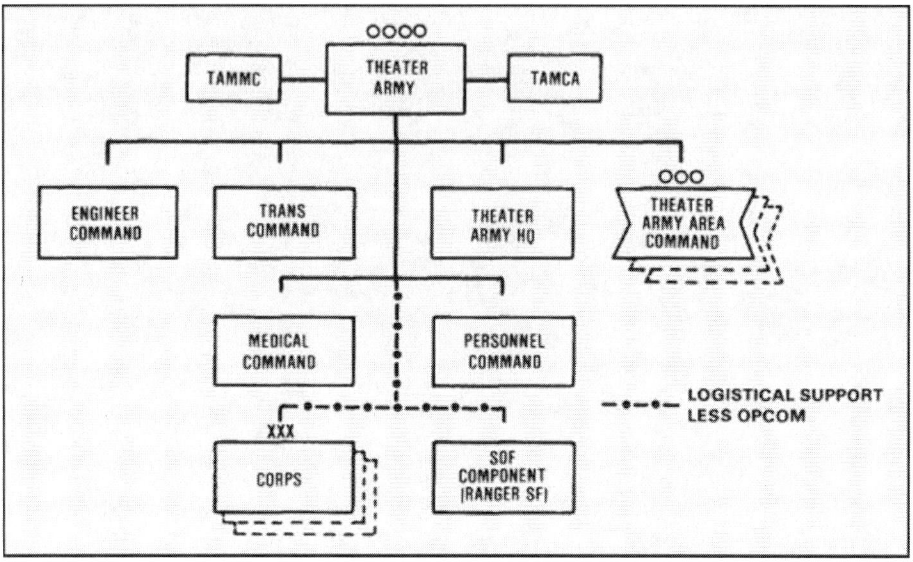

Figure 8-1. Ranger force relationship to theater Army.

c. The Theater Army Area Command (TAACOM) is the key CSS operator in the communications zone. It functions as a major subordinate command under the theater Army. The primary mission of the TAACOM is to provide direct support CSS to units such as the ranger regiment located in or passing through its assigned area. While located in an OCONUS REMAB, the ranger regiment receives most of its logistical support from the TAACOM.

d. The TAACOM area support group (ASG), which is subordinate to the TAACOM, is structured with operating units to provide direct support CSS to units located in or passing through its assigned geographical area of responsibility (see Figure 8-2). Area support groups are usually located along the lines of communication. The ASG is the prime source of CSS for the ranger regiment. It contains the units that actually provide services requested by the ranger unit logistics officer.

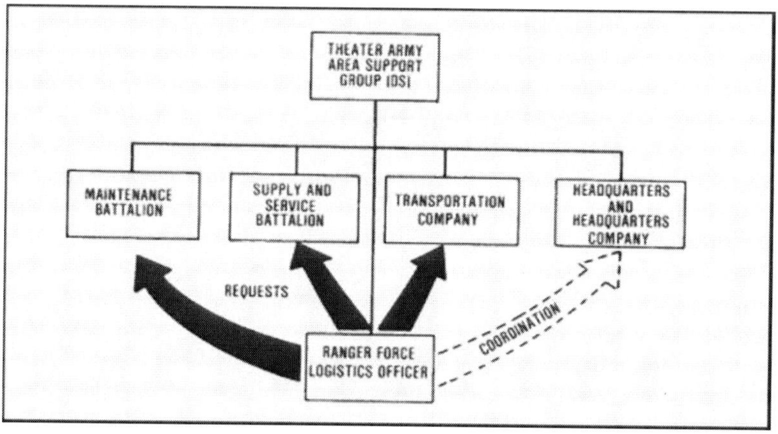

Figure 8-2. Ranger force relationship to theater Army area support group.

e. The ranger regiment is not normally assigned to a corps for employment. However, it may be placed under a corps' operational command for a specific mission. The ranger unit logistics and personnel officers must coordinate with the COSCOM that would then be providing CSS. (See Figure 8-3.) Corps support groups are major subordinate elements of the COSCOM. They provide command and control of COSCOM companies and battalions, which provide direct support (DS) and general support (GS) supply and field services to nondivisional units, such as the ranger battalion. (See Figure 8-4.)

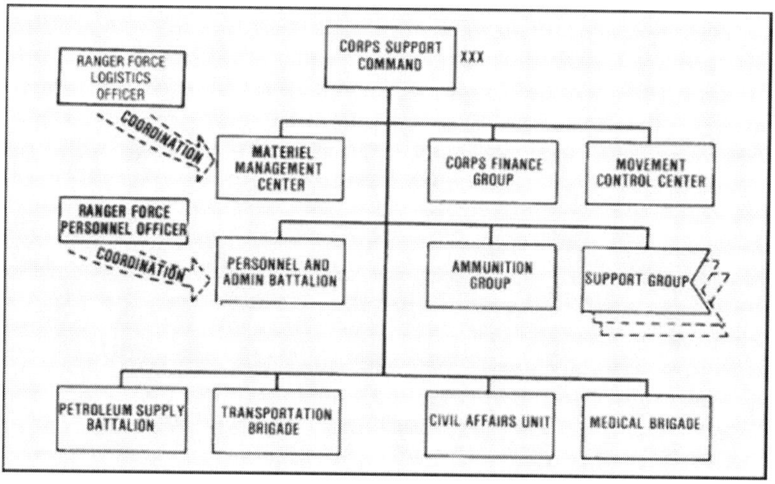

Figure 8-3. Ranger force relationship to corps support command.

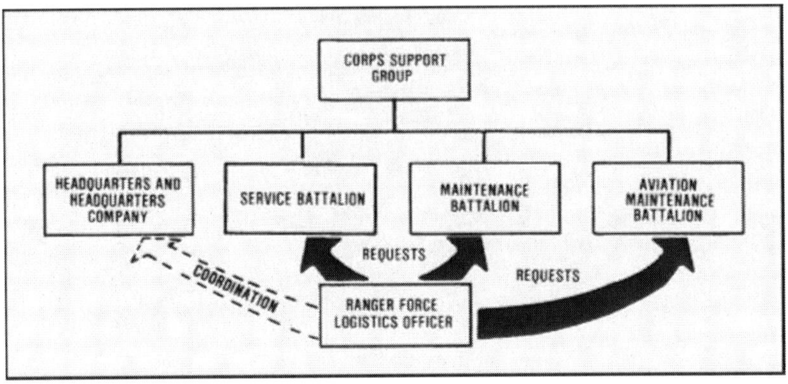

Figure 8-4. Ranger force relationship to corps support group.

f. Normally, the ranger regiment receives its logistical support from the corps support group responsible for the geographical area in which the REMAB or ISB is located. The corps support group provides all supplies except Classes V and VIII. The medical brigade provides Class VIII supply and medical maintenance support. The ammunition group provides Class V support. Financial and personnel services are provided by the corps finance group and the personnel and administration battalion, respectively.

g. Transportation needs are met by the transportation brigade. The ranger unit places ground transportation requests with either the Movement Control Center (MCC) or a designated movement control team. All airlift supply requirements are passed IAW FM 100-27. Standard and nonstandard supplies are normally requested from the supporting direct support unit (DSU). If authorized by the corps commander, the ranger regiment can place requirements for nonstandard items on the appropriate division within the corps MMC.

h. The COSCOM MMC (see Figure 8-5) performs supply and maintenance management for all classes of supply. The ranger unit logistics officer may coordinate with the MMC division chiefs or operations section.

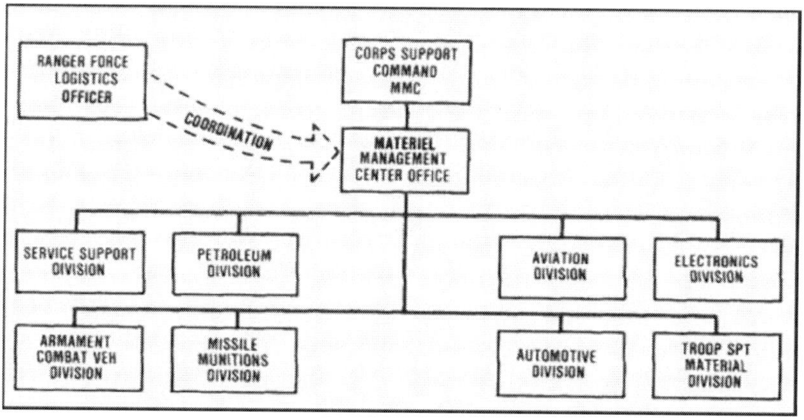

Figure 8-5. Ranger force relationship to COSCOM MMC.

i. Critical logistics operations, such as those supporting ranger and other SOF elements, need intensive management. They may be assigned to special staff sections that act as special operations force logistics managers. If designated, a logistics manager for special operations force elements would be a major point of contact for logistical support for the ranger regiment.

j. The ranger regiment may also receive logistical support from the SOSB. Normally, the SOSB provides administrative and logistical support to the Headquarters, Army Special Operations Forces (ARSOF). It complements the support provided by the Army component in the theater of operations by coordinating, augmenting, or assisting CSS forces. The SOSB provides unique quick-reaction support to the ARSOF or deploying SOF elements.

k. The ranger regiment does not make inordinate demands on the logistical system. However, a ranger unit may need certain nonstandard or low-density equipment. The concurrent need for nonstandard Class IX items requires management by logisticians to ensure support.

(1) Ranger battalion mess sections are responsible for Class I operations at the REMAB. The ranger regimental headquarters, which has no mess section, either collocates with one of the battalions or is supported by a mess facility operated by a nearby unit. It may share this facility on an area basis. The ranger unit personnel officer coordinates with the supporting DSU, which initiates the issue of Class I supplies.

(2) Class II, IV, and IX support to the ranger force is normally not of a great volume. It may consist of some nonstandard items. Depending on the need, the item, and the security, the SOSB may get some nonstandard items directly.

(3) Class III support needed by the ranger regiment is small and limited to standard items. The normal source of supply for Class III is the supply and service company providing area support.

(4) Class V support needed by the ranger regiment is small. It may include requests for nonstandard items of ammunition and explosives. The ranger logistics officer passes requests for standard items to the ordnance company (conventional ammunition) of the ammunition battalion assigned to the TAACOM ammunition group. Requests for nonstocked items are passed through the TAACOM Materiel Management Center or 1st SOCOM channels. The ranger unit must contact the ammunition company early to ensure a continuous supply of ammunition.

(5) Class VII supplies needed for the ranger regiment are often nonstandard or in excess of TOE. (See Appendix C for an equipment density listing.) All nonstandard or excess requests must be coordinated through the TAMMC or 1st SOCOM channels.

(6) Class VIII supplies for the ranger regiment are normally standard items supplied by the area support Class VIII point. The ranger regiment personnel officer sets up the Class VIII supply points of contact through the ranger battalion's medical officer.

(7) Aerial delivery equipment is provided by the airdrop equipment repair and supply company of the theater Army support group (TASG) (GS). There is normally one such company for each TAACOM. This unit supplies and repairs aerial delivery equipment. It does not pack and rig supplies for aerial delivery. The ranger regiment's requirements for aerial delivery supplies normally consist of standard items. Certain low-density items, such as HALO or HAHO parachute equipment, may also be provided through 1st SOCOM channels.

(8) The supporting motor transportation group of the transportation command is responsible for unit movement of the ranger regiment. The ranger regiment does not have any organic transportation assets. It depends on outside support whenever motor movement is needed.

8-3. Resupply.

a. Resupply operations for ranger units are normally covered during the planning and coordination phase of the mission. Ranger units try to complete their mission quickly and then be extracted or exfiltrate from the objective area. Ranger units carry into the objective area the equipment and munitions they need. They do not normally remain long enough to need extensive resupply. Certain operations may require the use of bulky or heavy items of equipment that cannot be carried by personnel. If the ranger force is airlanded on or near the objective, one or more of the aircraft may contain palletized supplies. These can be quickly off-loaded for future use.

b. If the ranger force is inserted into the objective area by parachute assault, aircraft can also drop resupply loads either just before or just after the personnel drop. These loads might include extra water, ammunition, demolitions, or barrier material. During peacetime, preloaded airdrop pallets are stored at CONUS depots for such contingency use. These mixed loads can be quickly moved to a departure airfield and inserted along with, or following, the ranger force. By dropping a resupply of water, food, equipment, and munitions with the ranger force insertion, the heavy-drop aircraft do not have to conduct a separate penetration of the enemy air defense zone. The ECM and SEAD programs used in the insertion of the personnel aircraft mask the penetration of the heavy-drop aircraft.

c. The aircraft dropping ranger personnel can also carry door bundles to deliver bulky or heavy items to the objective area. These door bundles, normally made up of A-7 and A-21 containers, may be color-coded to identify different loads or units. The most common door bundle loads are such items as 90-mm recoilless rifles and ammunition, 60-mm mortar ammunition, Dragon missiles, SHORAD missiles, or more demolition material. These bundles are pushed from the personnel aircraft just before the lead jumper's exit. They may be marked with lights or infrared sources for locating them at night.

d. If the tactical situation requires that a ranger force receive a resupply during the course of an operation, the most common method would be by airdrop. Airdrop could be needed if the operation takes longer than planned, initial supplies are lost, or changes occur in the operational plan. The ranger unit logistics officer and operations officers normally plan for resupply needs and schedule follow up loads of Classes I, II, IV, V, and VIII. There are two different airdrop requests channels for the ranger force to use:

(1) Nonemergency airdrop requests are sent through logistics channels. The ranger LOC sends requests for the airdrop of supplies and equipment to its supporting supply unit, and the supporting unit forwards the requests to the COSCOM or TAACOM MMC. The MMC coordinates with the movement control center (MCC). Jointly they direct the actions of the supply and transportation units. The MCC sends the request to the joint transportation board (JTB). The board validates it and assigns it a priority. The JTB then sends the request to the airlift control center for action. If the items requested are not on hand in the airdrop company, the MMC directs the proper supply activity to prepare the items for shipment to the airdrop unit. The MMC coordinates with the MCC, and the MCC ships the items. The airdrop unit prepares the supplies for the mode of delivery to be used. It also makes sure the supplies are delivered to the air terminal at the right time. (See Figure 8-6.)

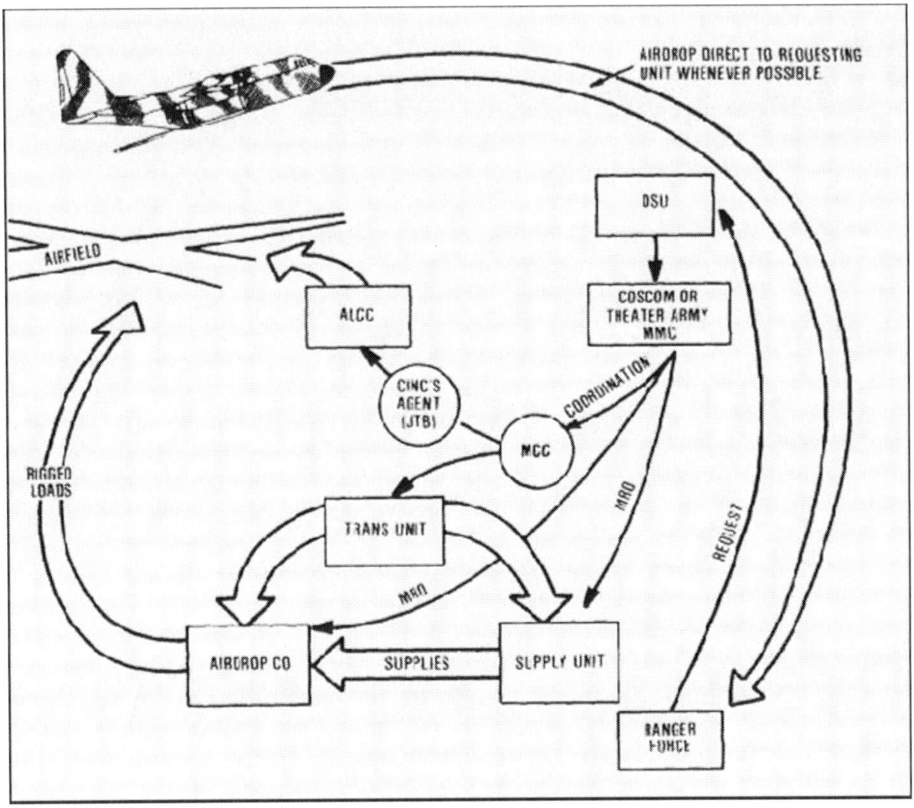

Figure 8-6. Ranger requests for nonemergency airdrop.

(2) Emergency requests for airdrop are passed through command channels in the quickest, most secure way.

(a) The request is usually sent to the command level that has authority to approve it, usually the tactical operations center at corps or above. The Air Force airlift officer at the direct air support center (DASC) coordinates with the Assistant Chief of Staff, Operations (G3) at the tactical operations center to determine whether airlift resources can be used for the mission.

(b) The TOC sends a request for an immediate airlift to the joint force commander's (JFC) designated agent for validation. At the same time, the airlift liaison officer advises the airlift control center (ALCC) that a request is coming through channels so plans can be made. When the ALCC receives the validated request from the JFC agent, it diverts or cancels less essential missions and provides airlift support. The TOC sends the request for supplies or equipment to the MMC.

(c) The MMC and MCC jointly direct the actions of the supply and transportation elements. The airdrop company stocks Classes I, III, and V supplies. It holds these supplies for emergency missions. If the supplies and equipment requested are not on hand in the airdrop company, the MMC directs the proper supply activity to prepare the items for shipment to the airdrop unit. The MMC coordinates with the MCC to ship the items. The airlift is arranged by the ALCC.

(d) The airdrop unit prepares the supplies according to the mode of delivery to be used. It coordinates with the MCC and delivers the supplies to the designated point at the air terminal (usually the cargo aircraft ramp). (See Figure 8-7.)

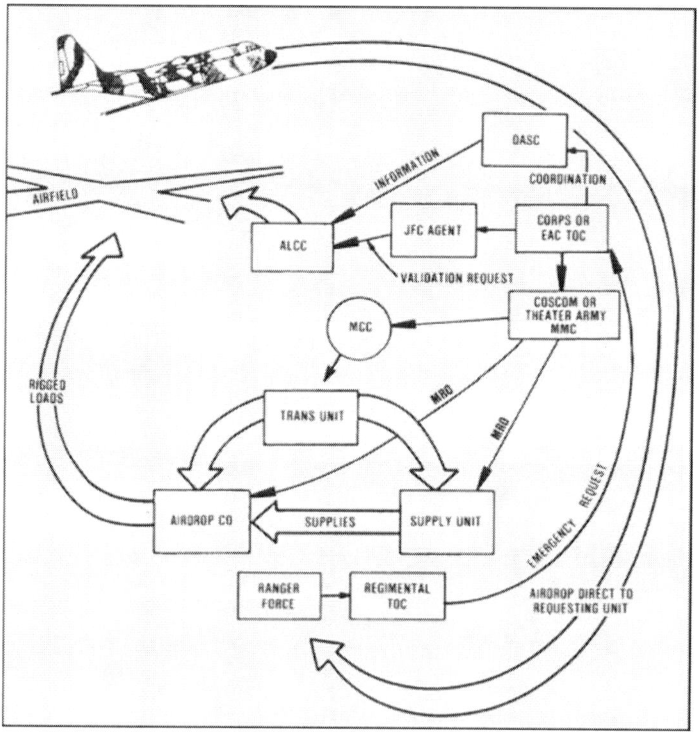

Figure 8-7. Request channels for emergency airdrop.

e. Aircraft delivering airdrop resupply deep behind enemy lines need to take precautions to avoid enemy detection and antiaircraft fire. The USAF is responsible for most of these precautions, such as ECM, suppression of enemy air defense (SEAD), and WILD WEASAL support. These do not concern the ranger commander. The choice of the type of parachute delivery system to use, however, may affect the ground commander's plan. The ranger force commander must specify any tactical restrictions that would require a specific type of delivery system. The ranger LOC and regimental logistics officer must advise the ranger force commander on the type of parachute system to be used. Conflicts must be quickly resolved so as not to interfere with the tactical operations.

(1) The safest way for the airdrop aircraft to penetrate enemy air defense and remain undetected is often by use of very low-level flight. There are four low-level parachute delivery systems that can be used without the aircraft being required to climb to a higher drop altitude and risk detection by enemy ADA.

(a) The low-altitude parachute extraction system (see Figure 8-8) needs a length of level ground with enough clearance for the delivery aircraft to make a very low pass over the extraction zone. The load is extracted by parachute from the aircraft and slides to a stop on a rugged, skid-type pallet. Heavy loads can be delivered using this system.

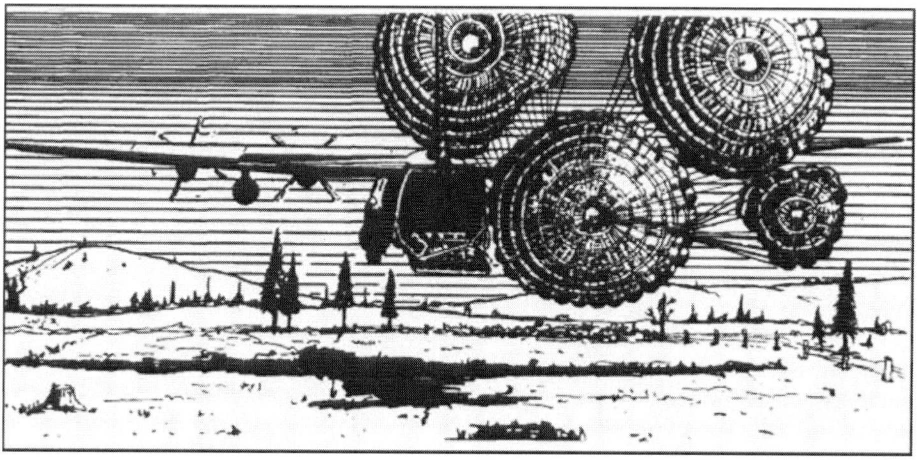

Figure 8-8. Low-altitude parachute extraction system.

(a) The containerized delivery system (CDS) (see Figure 8-9) consists of multiple, individually rigged A-22 containers, each with its own parachute, weighing up to 2,000 pounds each. The system provides single-pass delivery of up to 16 containers by C-130 aircraft and up to 28 containers by C-141 aircraft. The CDS loads may be delivered into drop zones using multiple points of impact to allow for tactical separation. The ranger force commander can direct the use of multiple points of impact, noting the advantages to be gained against the requirement for multiple passes.

Figure 8-9. Containerized delivery system.

(c) The high-speed, low-level, airdrop system (HSLLADS) (see Figure 8-10) is a single A-21 container specially rigged to withstand the parachute opening shock when airdropped from C-130 aircraft at high speed. This system can be used to deliver up to 600 pounds for each container, with a maximum of four containers for each pass, at speeds up to 250 knots.

Figure 8-10. High-speed, low-level, airdrop system.

(d) The CTU-2/A is a high-speed, aerial-delivery container (see Figure 8-11) that can be used to deliver supplies from high-performance aircraft flying at a minimum altitude of 300 feet above ground level and a maximum airspeed of 425 knots. The CTU-2/A is carried on the bomb racks of fighter or bomber-type aircraft. Upon release, a pilot parachute deploys the main chute and the container descends slowly.

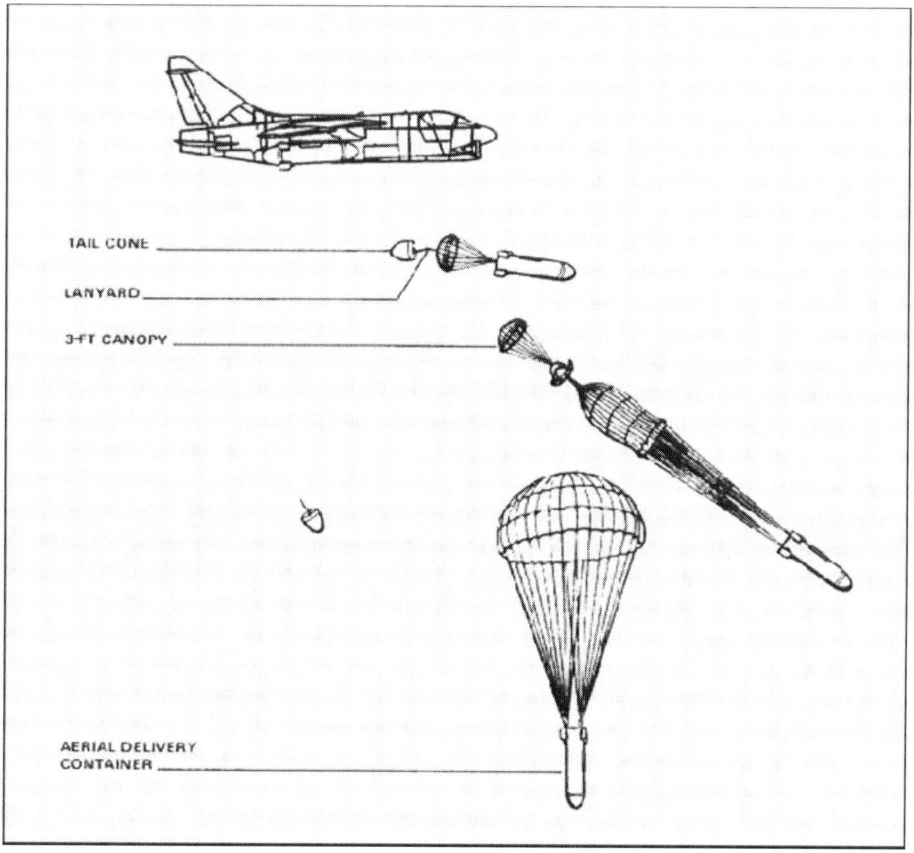

Figure 8-11. CTU-2/A high-speed, aerial-delivery container.

The CTU-2/A can be used to deliver up to 500 pounds of supplies such as weapons, water, food, or munitions. The container can be destroyed by burning. The main advantage of this system is that it can be delivered by high-performance aircraft deep behind enemy lines and in a dense air defense environment. The accuracy of this system is equal to that of a conventional bomb strike.

(2) Sometimes it may be better to drop resupply loads to the ranger force from a high altitude. The HAARS (see Figure 8-12) permits containerized unit loads weighing from 200 to 2,000 pounds to be delivered from aircraft at speeds up to 150 knots from up to 25,000 feet AGL. The HAARS consists of a cargo parachute, an airdrop container, an altitude

sensor, and a pilot chute. The pilot chute gives the descending bundle a speed slightly greater than an accompanying HALO parachutist. The system provides for the steady free-fall descent of loads from altitudes between 2,000 and 25,000 feet to an altitude at which a barometric sensor actuates deployment of the main parachute. This allows safe and accurate delivery of loads onto unprepared drop zones. The HAARS can deliver a payload to within 260 meters from a target impact point from a 10,000-foot altitude, with a proportional degree of accuracy from 25,000 feet. Ranger units can use the system for resupply of battalions and smaller units with rations, ammunitions, and medical supplies, breaking down the containerized material into man-packed loads.

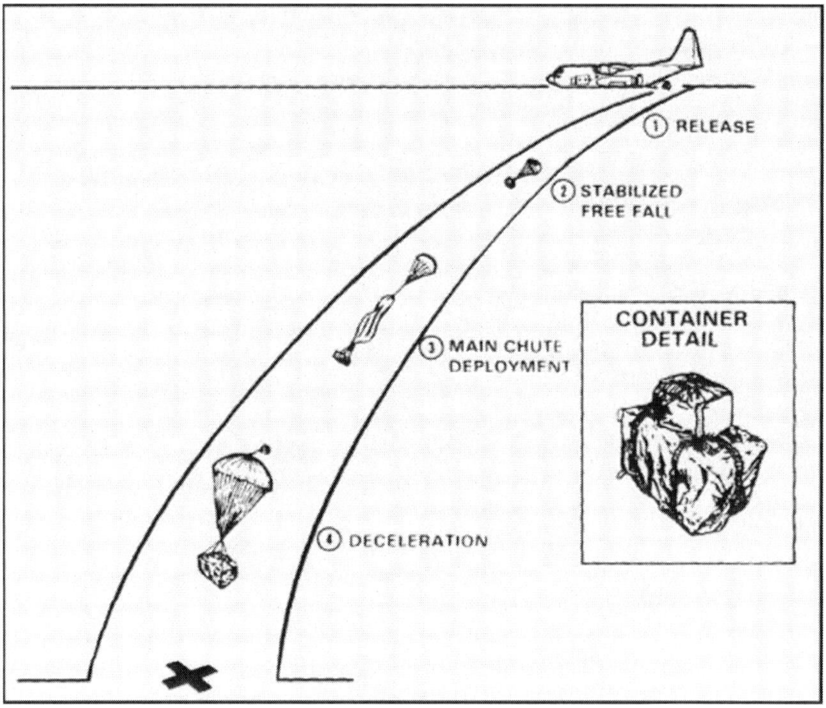

Figure 8-12. High-altitude, airdrop resupply system.

(3) The ranger force may need aircraft loads delivered during periods of poor visibility. The adverse weather aerial delivery system allows aircraft to drop resupply bundles accurately in adverse weather or darkness. These AWADS operations can be conducted safely and effectively with a minimum 300-foot ceiling AGL and a minimum visibility of less than half a mile. The minimum-drop altitude is 500 feet above the highest obstruction. These minimum ceiling or visibility restrictions do not apply to the Combat Talon aircraft.

f. Resupply to the ranger force can be provided by Army helicopters (see Figure 8-13). Normally this is done by SOA units. The danger to helicopters during long-range, low-level penetration of enemy rear areas requires that resupply be conducted when there is limited visibility.

Figure 8-13. Resupply bt helicopter.

8-4. Military police.

a. The ranger regiment needs military police support to provide security at the REMAB. Although ranger units can provide their own security and local defense, using military police for this purpose allows the ranger unit to concentrate its manpower and training on mission preparation. A military police security company is normally designated to provide support to 1st SOCOM elements located in the communications zone. Military police provide security patrols of the REMAB, convoy security, and vehicular search. They provide entry and exit control of the sensitive compartmented information facility (SCIF).

b. A preemptive strike by enemy special operations forces is a threat to the ranger regiment while located in the REMAB; either by conventional attack or by terrorist-type bombing. The preparation and location of the headquarters, the type of buildings, the location of landing zones, the traffic flow, and the size of the local population all influence the size of the military police unit. A military police platoon, augmented with bomb and guard dog teams, should be enough security at the REMAB. These security forces would not normally deploy from the REMAB to an ISB with the ranger unit.

8-5. Counterintelligence.

Counterintelligence support to the ranger regiment is normally provided by a CI team from the military intelligence group at TA level. The CI team provides a liaison and informational flow from the CI agencies operating within the theater. It also acts as advisor to the ranger unit commander on OPSEC while performing CI surveillance operations in the vicinity of the REMAB. The CI support may also be provided through 1st SOCOM channels. It is through 1st SOCOM that liaison with national intelligence and counterintelligence agencies is achieved.

8-6. Personnel service support.

The personnel service support needs of a ranger unit are much like that of any other unit of the same size. To receive the support his unit needs, the ranger personnel officer must meet with the proper agencies within the theater of operations.

a. **Strength accounting.** Within the ranger regiment, strength accounting is conducted through normal personnel channels. Regimental strength figures are passed quickly to the TAACOM or COSCOM. These figures are crucial to the theater commander for planning future operations. When matched with casualty reports, they provide the information needed for replacements.

b. **Replacement operations.** Ranger units need exceptionally well-trained and physically conditioned personnel as replacements. Ranger operations require that replacements train with the unit before they can become an effective member of the team. There are five sources of replacements for ranger units:

(1) **In-theater volunteers.** Use of in-theater volunteers, preferably with experience in ranger units, is the quickest way to replace personnel. These volunteers have a knowledge of the basic skills required of rangers and have some operational experience. The ranger regimental personnel officer coordinates with the theater personnel officer to screen and select those volunteers with the required MOSs and grades. Volunteers from other special operations force components, or from other theaters of war, are identified by DA and screened for assignment to the ranger regiment. Even though these volunteers have individual training and experience, it takes several weeks to become fully integrated into the ranger unit's operations and to achieve the needed skill level.

(2) **Reorganization of units.** This method of reconstitution is used when two or more ranger units have high personnel losses. The remaining members of these units are reorganized into a single unit. This method provides the ranger regimental commander with enough ready-trained, physically fit personnel to form a full-strength unit. Extensive unit training must be conducted before such a reorganization regains the unity and esprit of its component parts. The high standards found among the subelements of the ranger regiment makes this a good method of reconstituting a ranger unit. The units that have lost all their personnel may be reconstituted later, using some other method, or left unreconstituted,

depending on the situation.

(3) **Hospital returnees.** These personnel, former members of the ranger regiment who were injured, evacuated, treated, and have now recovered, are valuable assets. They provide the unit with a major reconstitution ability. These personnel should be returned quickly to their original battalion. This enhances unity and increases morale and esprit. Returnees may not be physically able to return to full duty right away. While recuperating, they may be assigned to limited duty as instructors. The ranger unit uses the returnee's combat experience and individual skills to screen and train newly assigned volunteers. Lessons learned in combat can help prepare the unit for future operations.

(4) **Replacement volunteers from CONUS.** These personnel have training in the basic skills and may have completed the ranger course, but they do not have any ranger unit experience. They need several weeks of unit training to become prepared for combat operations.

(5) **Ranger indoctrination program.** During wartime, this program provides screening and training of newly assigned personnel. Volunteers who have neither served in a ranger unit before nor have graduated from the ranger course require extensive individual training before they are ready to begin unit training with a ranger battalion. This method of reconstitution of ranger units requires the longest training period. During this reconstitution phase, a ranger battalion needs extensive use of ranges, maneuver areas, and support by aviation and other special operations-type units. The ranger regimental commander may choose to move the program from CONUS to a location near the regiment's REMAB.

c. **Casualty reporting.** Casualty reports are handled through regimental personnel channels and must be timely and accurate. The reports are used to check strength accounting data, to assist graves registration personnel in the recovery and identification of remains, and to notify next of kin.

d. **Health service.** The health service support needed by the ranger units deployed OCONUS is usually small. Area medical support provided by host country or theater medical command assets usually meets their needs. Evacuation of wounded members of the ranger force begins upon extraction from the objective area. Personnel wounded in action are evacuated through USAF or Army channels to the nearest medical facility that can provide treatment. The ranger force medical officer and personnel officer coordinate with the area medical support facility providing preventive and routine medical care to the unit while at the REMAB. Vaccine needed for innoculation of the ranger force members and any special Class VIII supplies needed for the mission are drawn through normal medical channels.

e. **Financial service.** Financial support varies with the situation. In mid- and high-intensity conflicts, regular paydays are suspended in the theater of operations. As a nondivisional unit, the ranger regiment coordinates with the theater finance center or corps finance group for financial support. In the REMAB, ranger units are provided financial support from the closest finance support unit. The ranger unit personnel officer appoints Class A agents to help with the disbursement of funds. The ranger force may be given a specific amount of local currency or military scrip depending on the length of time in the REMAB and the operational situation. The ranger unit personnel officer also coordinates with the theater finance center to get any third-country currency that may be included as part of an escape-and-evasion kit provided to ranger elements.

f. **Postal service.** Postal service to the ranger unit is through normal channels. Mail for members of the deployed force is forwarded by the rear detachment commander to an established theater Army general support postal activity. This activity forwards the mail to the postal unit designated to support the REMAB site. The ranger unit commander appoints subordinate unit officers as censors. The exact location of the REMAB, and all aspects of the mission and its location, remains classified until released by an official Army spokesman. The ranger unit commander must ensure that the forwarding of mail, or the writing home of personal letters, does not endanger the ranger force mission. All outgoing mail is censored during wartime. For security reasons, outgoing mail may be held for a short time before it is forwarded. Mail from the ranger unit is never sent by host country or other nation's postal services.

g. **Chaplain activities.** Chaplain activities are provided by the chaplain assigned to the ranger battalion. He provides unit coverage through services, rites, and sacraments regardless of his or the unit member's religious affiliation.

h. **Legal service.** Ranger units do not need extensive legal service support. Legal service support is obtained through coordination with the theater Army staff judge advocate.

APPENDIX A

Organization and Equipment Introduction of the Ranger Regiment

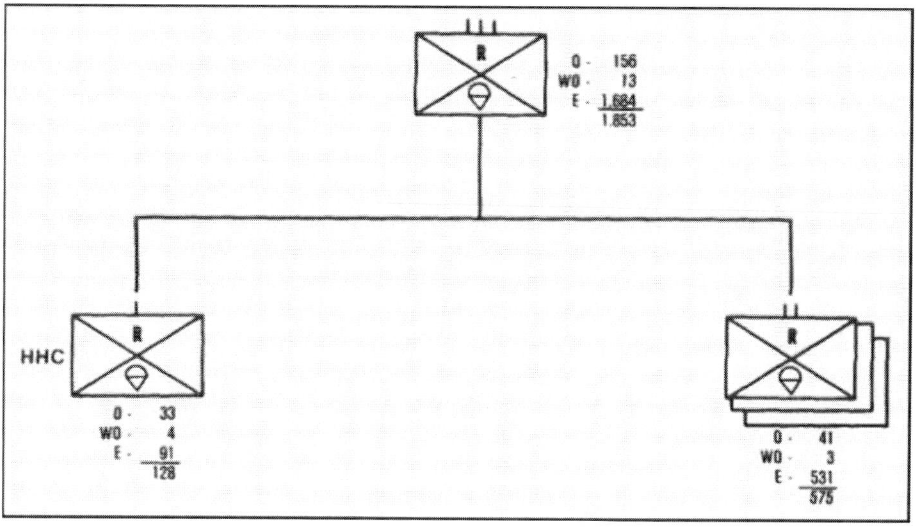

Figure A-1. Ranger regiment.

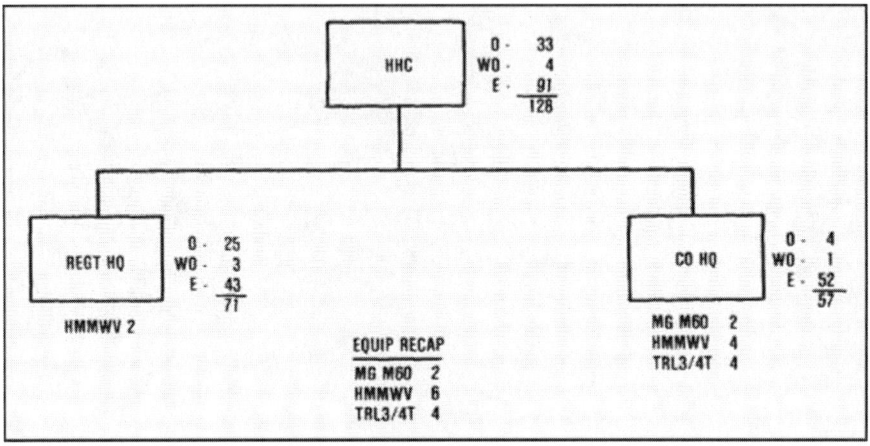

Figure A-2. Ranger regimental HHC.

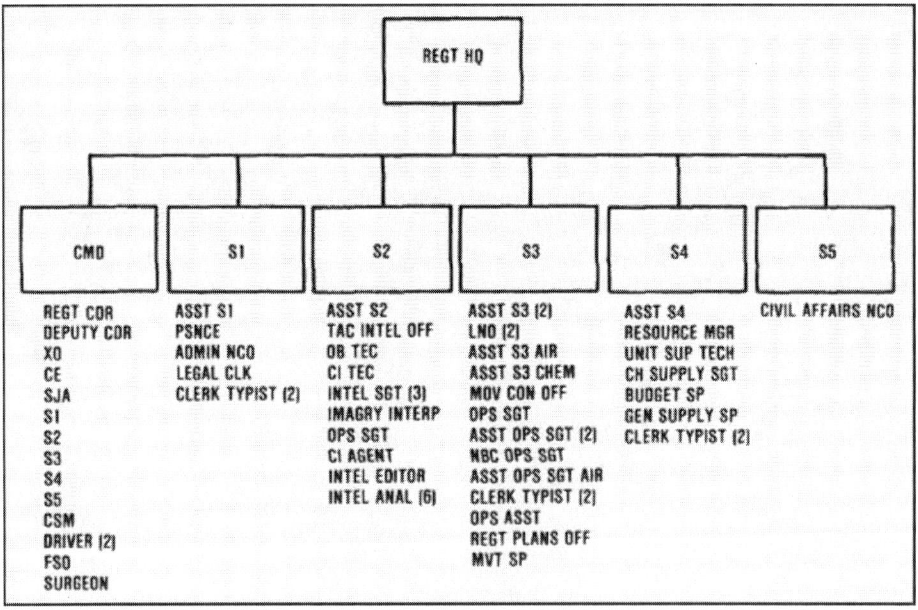

Figure A-3. Ranger regimental staff.

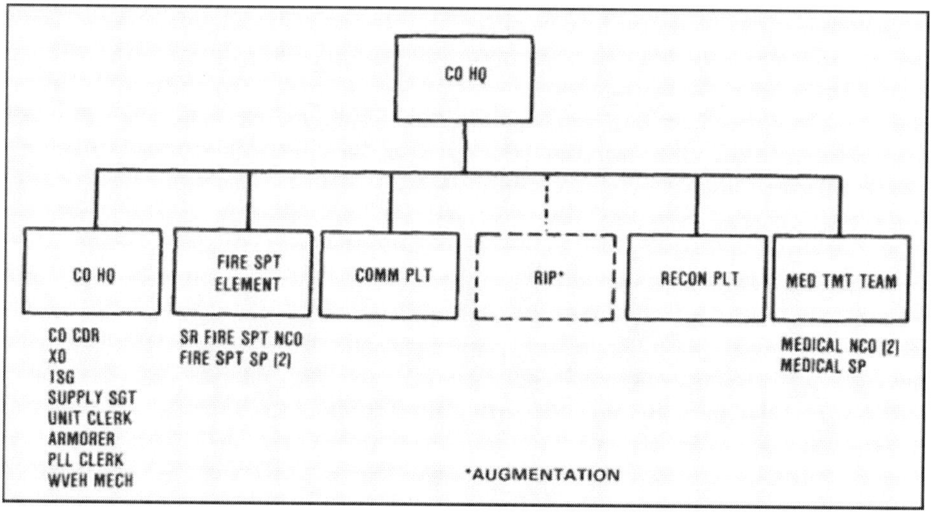

Figure A-4. Ranger regimental headquarters company.

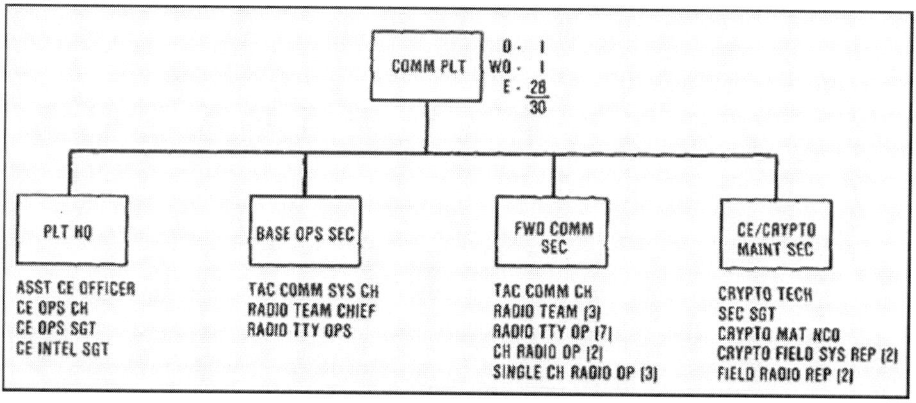

Figure A-5. Ranger regimental communications platoon.

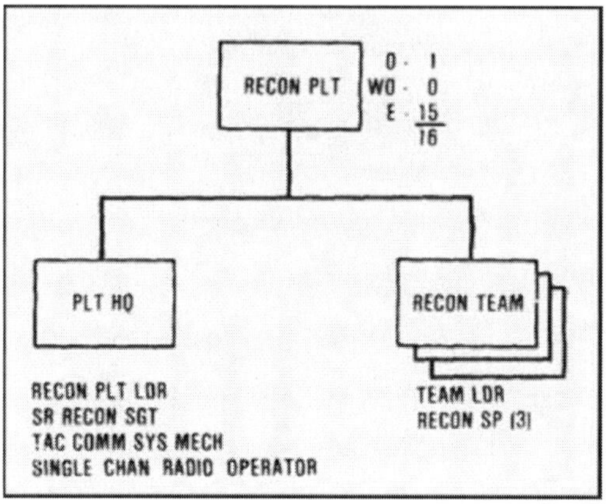

Figure A-6. Ranger regimental reconnaissance platoon.

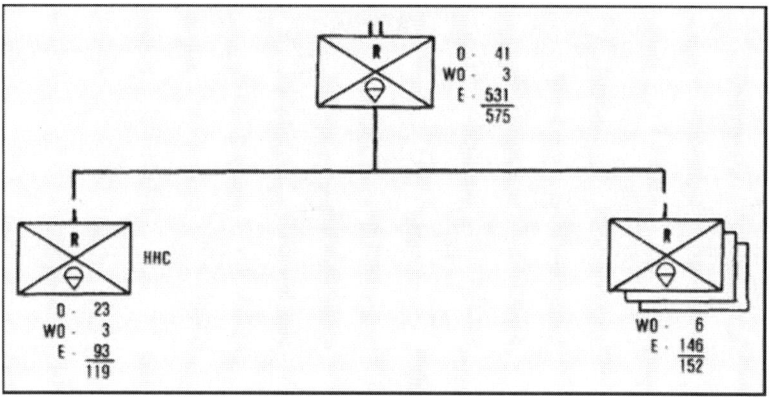

Figure A-7. The ranger battalion.

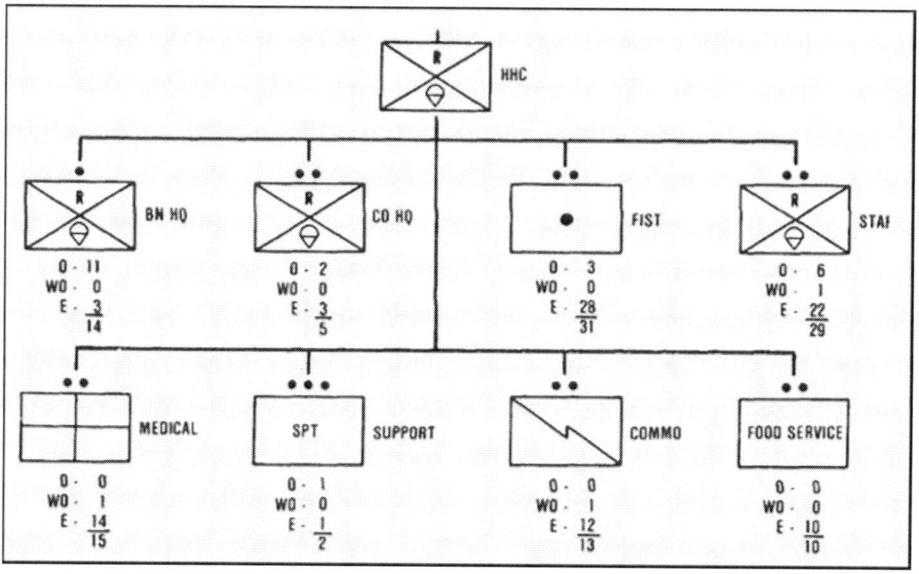

Figure A-8. Ranger battalion HHC.

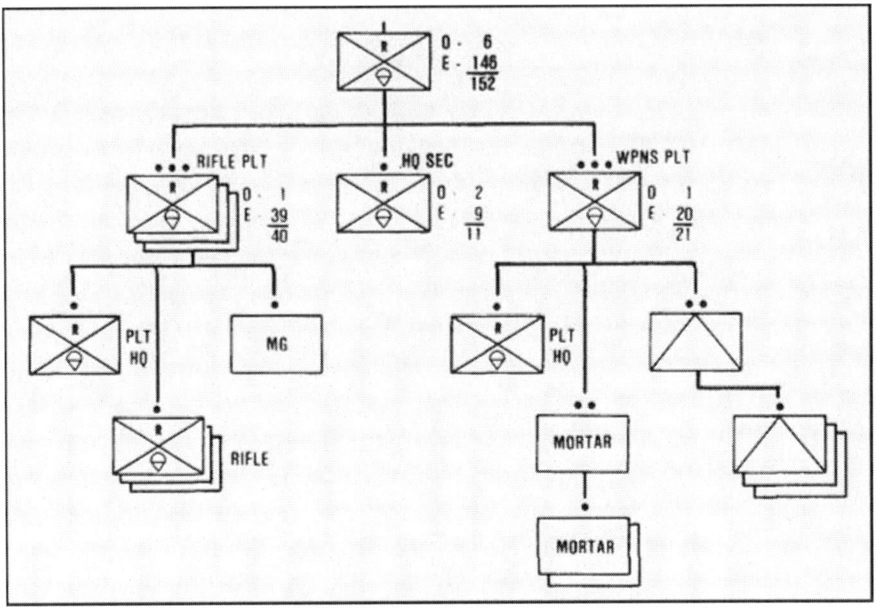

Figure A-9. Ranger rifle company.

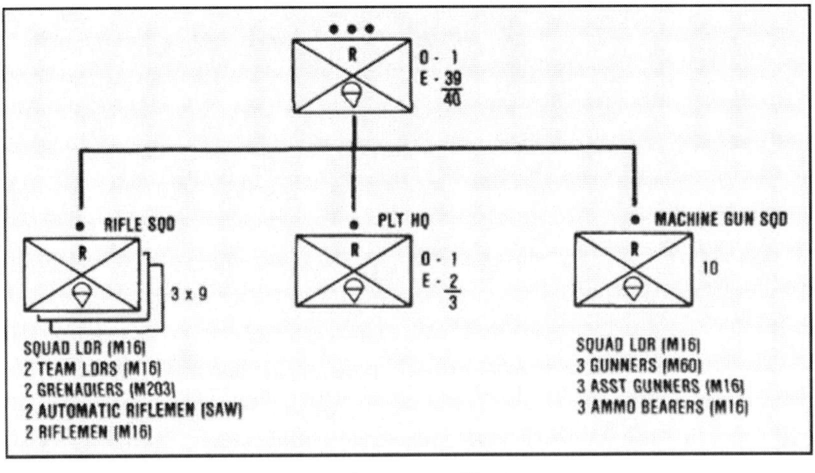

Figure A-10. Ranger rifle platoon.

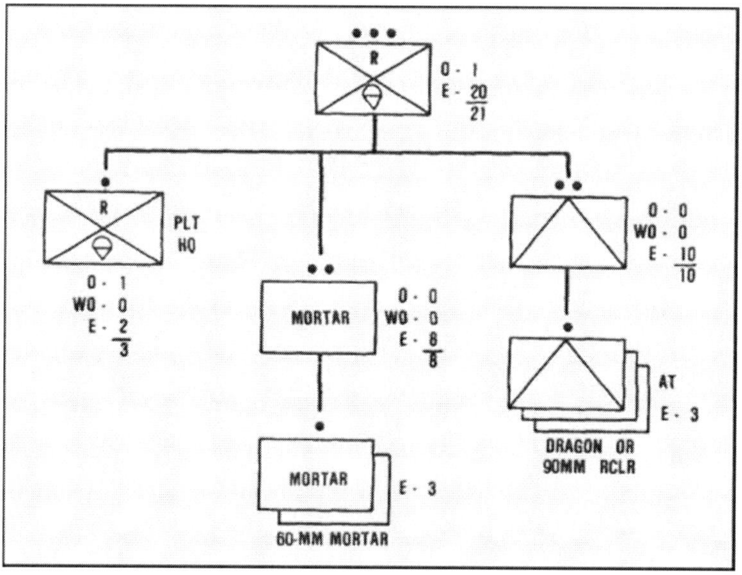

Figure A-11. Ranger weapons platoon.

APPENDIX B

Composition and Equipment of Ranger Regiment Liaison Teams

Team Red	**RANK**
Liaison Officer	MAJ
Senior Intelligence Analyst	SSG
Clerk-Typist	SP4
Tactical Communications System Operator/Mechanic	SGT
Combat Telecommunications Center Operator	SP4
Tactical Communications System Operator/Mechanic	SP4
Team White	
Liaison Officer	CPT
Senior Intelligence Analyst	SSG
Clerk-Typist	SP4
Tactical Communications System Operator/Mechanic	SGT
Combat Telecommunications Center Operator	SP4
Tactical Communications System Operator/Mechanic	SP4

Equipment (for Each Team)

HMMWV with 3/4-ton trailer	5-kw generator
Communications shelter	Tent, GP, small
AN/PCS-3 radio set (3)	AN/VSC-7 net control station (1)
MX 300 (4)	KY 57 with battery case (5)
HYX 57 with battery case (4)	HYP 57 (5)

KYK 13 (2)

KY 65 with battery case (1)

KG 84 (1)

H189 (2)

GRC 193 (1)

VGC-74 (1)

PP 1648 (1)

SB-22 switchboard (1)

KYI 18 (1)

KY 65 WLA (1)

Digital message device (1)

AN/PRC-77 radio (1)

Facsimile transmission device (1)

TA 312 with WD1 (5)

APPENDIX C

Equipment Density

C-1. Communication equipment.

	Regimental Headquarters	Battalions (x3)
AN/VSC-7 net control station	7	2
AN/PSC-3 radio set	23	13
AN/PRC-41 radio set	2	5
AN/PRC-104 radio set	9	26
AN/GRC-193A radio set	4	0
AN/PRC-70 radio set	6	0
AN/PRC-77 radio set	22	50
AN/VRC-46 radio set	2	0
AN/VRC-49 radio set	0	2
AN/PRC-68 radio set	0	66
TSEC/KY-57 speech security device	60	68
TSEC/KY-58 speech security device	6	0
TSEC/KY-65 speech security device	11	0
AO-8990/P digital message device group	15	23
AN/GXC-7A tactical FAX device	7	2
AN/UGC-74A(V)3 communications terminal	7	0
SB-22 switchboard	2	0
SB-3614(V)/TT switchboard	2	1

TSEC/KG-84	5	0
SB-993/GT switchboard	0	12

C-2. Weapons and equipment each battalion.

Infrared aiming light	96	Interrogator set, Stinger missile	3
Rifle, M16	451	Grenade launcher, M203	67
Rocket launcher, M202	13	Machine gun, M60	27
Mortar, 60-mm M224	6	Rifle, sniper, M21	24
Recoilless rifle, 90-mm	18	Viewer, infrared	9
Tracker, Dragon	9	Night vision devices (all types)	322
Truck, 1/4-ton	2	Laser target designators	11
Radio (all types)	164	Tactical satellite radios (all types)	15
Squad automatic weapon	54		
Laser range finder	14		

APPENDIX D

Standing Orders, Rogers' Rangers

1. Don't forget nothing.
2. Have your musket clean as a whistle, hatchet scoured, sixty rounds powder and ball, and be ready to march at a minute's warning.
3. When you're on the march, act the way you would if you was sneaking up on a deer. See the enemy first.
4. Tell the truth about what you see and what you do. There is an army depending on us for correct information. You can lie all you please when you tell other folks about the Rangers, but don't never lie to a Ranger or officer.
5. Don't never take a chance you don't have to.
6. When we're on the march we march single file, far enough apart so one shot can't go through two men.
7. If we strike swamps, or soft ground, we spread out abreast, so it's hard to track us.
8. When we march, we keep moving till dark, so as to give the enemy the least possible chance at us.
9. When we camp, half the party stays awake while the other half sleeps.
10. If we take prisoners, we keep 'em separate till we have had time to examine them, so they can't cook up a story between 'em.
11. Don't ever march home the same way. Take a different route so you won't be ambushed.
12. No matter whether we travel in big parties or little ones, each party has to keep a scout 20 yards ahead, 20 yards on each flank, and 20 yards in the rear so the main body can't be surprised and wiped out.
13. Every night you'll be told where to meet if surrounded by a superior force.
14. Don't sit down to eat without posting sentries.
15. Don't sleep beyond dawn. Dawn's when the French and Indians attack.
16. Don't cross a river by a regular ford.

17. If somebody's trailing you, make a circle, come back onto your own tracks, and ambush the folks that aim to ambush you.

18. Don't stand up when the enemy's coming against you. Kneel down, lie down, hide behind a tree.

19. Let the enemy come till he's almost close enough to touch. Then let him have it and jump out and finish him up with your hatchet.

MAJ Robert Rogers

APPENDIX E

The Ranger Creed

R Recognizing that I volunteered as a Ranger, fully knowing the hazards of my chosen profession, I will always endeavor to uphold the prestige, honor, and high "esprit de corps" of the Ranger Regiment.

A Acknowledging the fact that a Ranger is a more elite soldier who arrives at the cutting edge of a battle by land, sea, or air, I accept the fact that as a Ranger, my country expects me to move farther, faster, and fight harder than any other soldier.

N Never shall I fail my comrades. I will always keep myself mentally alert, physically strong, and morally straight and I will shoulder more than my share of the task, whatever it may be. One hundred percent and then some.

G Gallantly will I show the world that I am a specially selected and well-trained soldier. My courtesy to superiors, my neatness of dress, and my care of equipment shall set the example for others to follow.

E Energetically will I meet the enemies of my country. I shall defeat them on the field of battle, for I am better trained and will fight with all my might. Surrender is not a Ranger word. I will never leave a fallen comrade to fall into the hands of the enemy and under no circumstances will I ever embarrass my country.

R Readily will I display the intestinal fortitude required to fight on to the Ranger objective and complete the mission, though I be the lone survivor.

RANGERS LEAD THE WAY

APPENDIX G

Training

G-1. Application and guidance.

a. To prepare for combat, ranger units train on missions derived from their analysis of operational and contingency plans and the ranger mission training plans (MTPs). Training guidance is in doctrinal manuals, field circulars, and soldier's training publications. Training must ensure that the ranger unit is skilled in all basic light infantry skills and in ranger-unique skills. In the final analysis, good training depends upon the leadership and teamwork of ranger unit officers and NCOs.

b. Everything rangers do should prepare them to perform well in combat. Ranger units train IAW the battalion training management system (BTMS) concentrating on basic skills and on live-fire, hands-on training.

c. An effective training program breeds high morale and few discipline-related problems. Some qualities of ranger units that have good training programs are as follows:

(1) Leaders who can solve problems.

(2) A thorough and continuous analysis of the unit's mission.

(3) Validated training objectives.

(4) Attention to the basics (move, shoot, and communicate).

(5) Multiechelon sustainment training.

(6) Standardized procedures.

(7) Physical and mental strength.

(8) Live fire as an integral part of training.

(9) Safety as a basis to all operations.

(10) The use of night and adverse weather and difficult terrain as combat force multipliers,

d. When all training is planned with realism, rangers will be in training situations that require the same attention as in battle. Realism and stress in training result in faster learning and better retention of skills. Training objectives must be clear with measurable standards to determine proficiency. Rangers and ranger units must train to the standards that will be expected in combat.

e. Examples of unit training goals include:

(1) Developing each member of the chain of command to perform his duties with competence, confidence, and pride.

(2) Maintaining small unit integrity during all training and support activities. Accomplishing tasks as units.

(3) Ensuring that all tasks for training are keyed to mission needs and weaknesses. Assets--ranger's time, money, and materiel--are expended only after thorough planning.

(4) Training to standard.

(5) Developing and maintaining discipline and esprit to ensure the abilities of each ranger and all units are achieved.

(6) Emphasizing use of the ranger mission training plan (MTP) for each echelon of the organization.

(7) Emphasizing the basics.

(8) Maintaining a high level of physical fitness.

(9) Working together within the ranger regiment and other elements of the special operations forces community.

(10) Developing a unit atmosphere of quick reaction under stress. Encouraging and rewarding individual initiative. Accomplishing unit missions IAW the commander's intent.

G-2. Philosophy.

a. To be effective, training should be physically and mentally challenging. Each ranger will then have a sense of purpose, accomplishment, and satisfaction. Training is evaluated by the results rather than hours of training. Units that are well trained today need reinforcement training to sustain readiness. Training management must be continuous. Training managers must use the time available to improve training efficiency. Cyclic training will not produce the desired results. The "train-to-correct deficiencies" method is needed. The training managers must analyze weaknesses, conduct prescriptive and reinforcement training, and reevaluate their unit's training status. The process that is used to develop and maintain a continuous training and evaluation program is shown in Figure G-1.

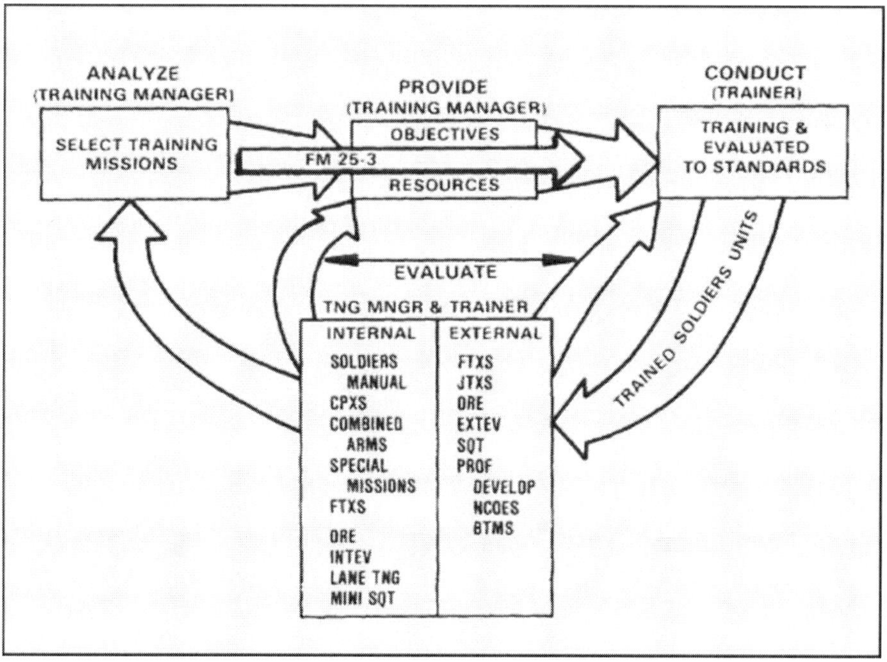

Figure G-1. The Army training and evaluation program.

b. Individual and collective training must be integrated within a unit training program. Training of the individual ranger, leaders, and units at each level must occur at the same time. The ranger commander or leader plans and conducts training for every member of his unit.

c. Ranger commanders analyze operational and contingency plans to determine unit missions and translate the associated tasks into training objectives. These objectives form the basis of the unit training program. The regimental headquarters provides training guidance in a formal regimental training plan. Commanders at lower levels program, supervise, and conduct training within their units.

d. All members of the chain of command are responsible for training. The NCO leader or first-line supervisor conducts individual and small unit collective task training. The officer plans individual training in his unit and directs and conducts collective training.

e. Centralized development of training objectives and decentralized training builds well-trained units and outstanding leaders. Authority and responsibility for the detailed planning, scheduling, conducting, evaluating, and supervising of training are delegated to the lowest element capable of managing this training. Higher headquarters is responsible for establishing objectives, monitoring and evaluating training, and issuing general guidelines.

f. Multiechelon training and evaluation means using all assets and time available to improve combat readiness. The multiechelon structure of the ranger MTP provides a basis for units to be continually trained and evaluated by internal evaluation. Each leader must know his mission tasks and train to accomplish those tasks to the set standards.

G-3. Mission analysis.

a. Mission analysis is continuous for a good training program. The selection of training missions or tasks results from analysis of the unit's overall mission. (See Figure G-1 above.) Tasks and subtasks are then placed in order of priority.

b. The trainer must identify countertasks for the opposing force (OPFOR). The countertasks should be such that an OPFOR that meets the standards for its countertask will be successful. Both sides have a chance to win.

c. The trainer should then develop combat drills and situational training exercises (STXs), if not furnished by the proponent school.

G-4. Unit training.

a. **Responsibility.** Ranger commanders at all levels are responsible for all training and management described herein and current Army training doctrine. This includes individual and collective training at all levels. Units must train to the standards specified in training objectives.

b. **Management.** The training management model for commanders is:

- Set objectives.
- Provide resources.
- Coach subordinates.
- Measure results.
- Provide feedback.

G-5. Individual training.

There are three elements basic to a well-trained individual ranger. He must be physically fit, be proficient in the use of his weapon, and have the psychological commitment to fight the enemy.

a. Individual training is the responsibility of the first-line supervisor under the supervision of the chain of command. It is conducted using performance-oriented techniques under battlefield conditions. New skills must be developed and practiced, and acquired skills must be reinforced throughout the training year. A high level of proficiency must be maintained through continuous training and practice.

b. Trainers should use peer instruction. Commanders ensure needed resources are available to conduct challenging individual training. Ranger units schedule and conduct this training and evaluate individual proficiency.

c. Flexibility in planning and scheduling is necessary. The goal of all individual training must provide for reinforcement training to offset the effects of the forgetting curve. (See Figure G-2).

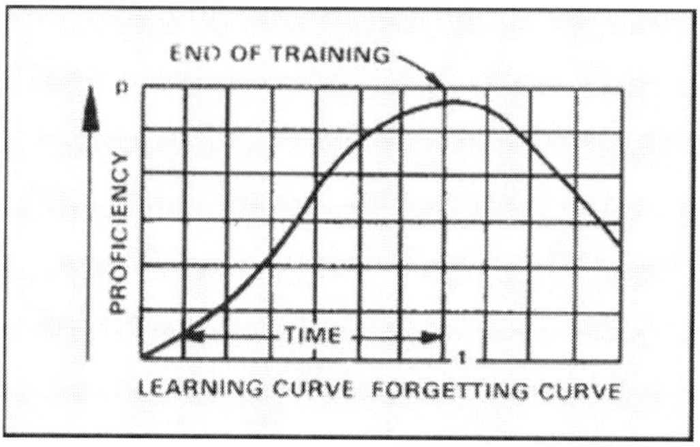

Figure G-2. The learning curve and forgetting curve.

d. The command sergeants major, unit first sergeants, and staff section sergeants are the keys to an effective training program. These senior NCOs must counsel, guide, and critique junior NCOs in training preparation and execution. Fire team leaders, section chiefs, and squad leaders coordinate and present individual training.

e. The NCO supervisors must ensure that the ranger is trained in the skills required for the specific military occupational specialty (MOS) and in those skills used daily. Not all training can be acquired in an office. For example, all rangers must maintain a high skill level with their weapons, be trained in NBC subjects, and be capable of performing contingency missions.

G-6. Collective training.

a. **Guidance.** Collective training trains crews, teams, squads, platoons, companies, and battalions to do their missions. The ranger MTP is used to plan training and evaluate the unit's weaknesses and strengths. If a ranger commander receives new training requirements or missions, they should become part of the collective training program. The ranger MTP describes the combat tasks to be performed by every echelon of the ranger battalion. It is a plan for organizing, conducting, and evaluating the year-round unit training program.

b. **Responsibility.** The ranger unit commander is responsible for the conduct of collective training and for all other training in his unit. Practical exercises emphasize the techniques of terrain use, movement, and effective employment of weapon systems. Training under stress will be conducted to move the unit to a higher plane of combat readiness.

c. **Drills.** Drills are written for small units--for example, fire team and squad, crew and section--and describe an action that is shorter than an entire STX. Drill training objectives cover both individual and collective skills (critical collective tasks) and list them together for effective group action. Drills must be mastered in order to react to rapidly changing situations in combat. The trainer can interrupt the execution of a drill to provide necessary training on individual or collective skills.

d. **Situational training exercise.**

(1) The STXs are short, mission-oriented tactical exercises in which a group of closely related collective tasks are trained. They are like drills but are more complex and flexible. They usually involve a larger unit than does a drill. They integrate leader training, leader tasks, drills, and separate individual tasks into a realistic setting. Trainers may execute STXs as written or tailor them to meet specific unit requirements. An STX is not standardized, but it must have a firm doctrinal base.

(2) Due to their design and purpose, STXs apply mainly to platoon- and company-size units. This does not preclude the development of STXs or similar exercises for unit staffs and headquarters. They combine "how-to-train" specifics for selected collective tasks with the "what-to-train" specifics found in doctrinal manuals and other documents.

G-7. Performance-oriented training.

a. Performance-oriented training emphasizes hands-on performance. Training is conducted to ensure a level of proficiency, rather than established number of hours of required training.

b. Performance-oriented training has precise training objectives and makes good use of resources. Training objectives serve as the basis for preparing, conducting, and evaluating all training.

G-8. Evaluation.

a. A training program must include an evaluation of training (planned and conducted). An evaluation plan must:

(1) Provide immediate feedback to the trainer.

(2) Recognize effective trainers.

(3) Evaluate all units in the conduct of training to provide for a comparison of training programs.

(4) Be flexible to unit needs; provide general and specific evaluations based on the level of training within the unit.

(5) Be in effect at all times.

b. Major considerations for training evaluations include:

(1) Ranger unit commanders establish and maintain an effective internal and external evaluation program.

(2) The director of each staff element is responsible for training evaluations within his area of staff responsibility with emphasis on low-density MOS groups.

(3) The operations officer of the ranger organization develops and validates training and evaluation plans.

c. Evaluation of training must be done vertically from the major task (mission) down through subtasks (collective tasks) to the leader and individual tasks. In selection of individual tasks for evaluation, priority should be given to critical individual tasks that impact on the success of the collective task or mission.

d. The critique and after-action review, after each major training event, are the most important aspects of training. The after-action review takes place immediately after completion of a training task (individual or collective). The individual(s) self-analyzes and critiques the training event. Then after a formal critique of the training, the individuals) trains to standard (not to time).

G-9. Field training.

a. **Considerations.** Ranger commanders schedule training under field conditions to permit maximum training opportunities. Field training emphasis is on mastery of a series of individual (soldier and leader) tasks and collective training tasks. These tasks are repeated until established standards are met. The goal of ranger field training is to ensure realistic, demanding, and meaningful training.

(1) When training is conducted under conditions of moderate stress, most combat critical skills are learned quicker and retained longer. (See Figure G-3 on the following page.) Stress can be incorporated in a training situation by physical deprivation (sleep, food), by live-fire exercises, by operation against an OPFOR under simulated battlefield conditions for an extended period, by the intensity of activity including demands for decision making, or by physically demanding activities. Ranger units train under stress so that they will react quickly in combat.

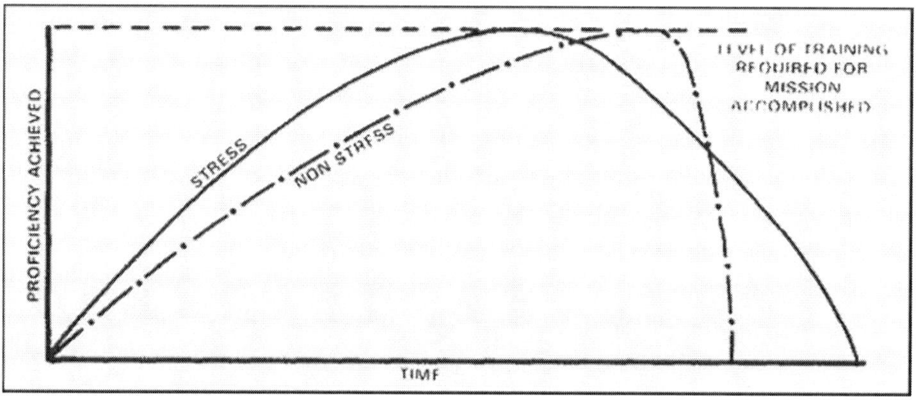

Figure G-3. The Army training and evaluation program.

(2) Antagonistic training situations are two-sided exercises conducted by opposing elements. Each element has a training objective (task and standards). The opposing force has a corresponding countertask. This training may be progressive--for example, starting with a

crawl-through, then a walk-through, and progressing to both elements operating at a fast pace. The use of multiple integrated laser equipment system (MILES) greatly enhances task and countertask training. The units should also be considered for a live-fire exercise against a realistic OPFOR target array when the basics have been mastered.

b. **Methods of training.**

(1) **Considerations.** The chance for success in combat can be increased by becoming proficient in training situations that depict realistic combat conditions. (See Figure G-4.) Meeting the standards of a realistic live-fire training exercise assures the commander that his unit could perform well in combat. Figure G-4 shows the relationship between realistic training and the challenge to the trainer.

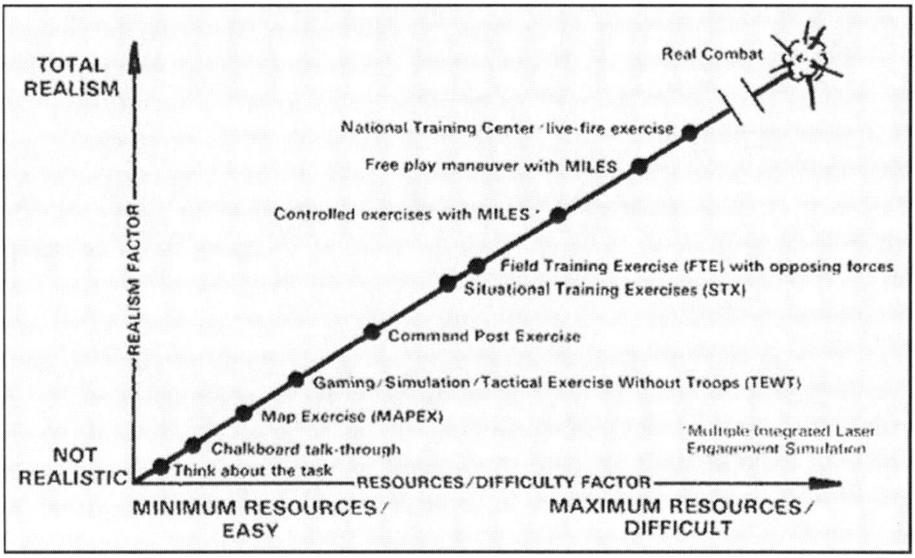

Figure G-4. Training proficiency scale.

(2) **Ranger cohesiveness training.** Ranger units promote teamwork by undergoing tough, stressful, often hazardous training. This challenges the unit, which gains in cohesiveness when it meets these challenges. This includes serving together in live-fire exercises, athletic events, precision parades, high-standard inspections, tough physical training, and combat.

(3) **Field training exercise.** An FTX is training normally conducted under simulated, realistic combat conditions. This training normally exercises command and control systems and the administrative and logistical needs of sustained combat. Stress training is part of this exercise, focusing on live-fire and MILES drills and STXS.

(4) **Combined arms training.** Combined arms training should include all means of support such as tactical air, armed helicopters, and indirect fire with emphasis on fire support coordination and airspace control.

(5) **Live-fire exercises.** Including live-fire exercises in drills and STXs offers a chance to evaluate units in their response to simulated combat missions. Ranger units train under pressure and always anticipate being selected to conduct a live-fire exercise. Live-fire exercises develop confidence and esprit, and reinforce unit discipline.

(6) **Opposing force.** Committing one unit against another adds a competitive spirit to a field training exercise (FTX). The use of MILES and squad combat operations exercises (simulation) (SCOPES) provides realism to two-sided field exercises. Trainers should use OPFOR equipment and training devices available, as well as making their own devices. (See FM 30-103, Appendix R, for more information.)

c. **Key functional training areas.**

(1) **Land navigation and patrolling.** Day and night navigation and patrolling exercises must be emphasized. Orienteering improves these skills, as well as physical fitness.

(2) **Movement techniques.** Infantry combat has two major components--to move and to hit the enemy first. Ranger units must be skilled in the use of traveling, traveling overmatch, and bounding overmatch. Units in contact must be skilled in moving under fire.

(3) **Fighting** positions. Training includes the building of concealed fighting positions.

(4) **Mine and countermine training.** Training includes how to defend against hostile mines, booby traps, and other demolitions. The identification, installation, neutralization, and removal of US mines and devices is also stressed.

(5) **Limited visibility operations.** Ranger units must be skilled in limited visibility operations and the use of night vision and thermal devices.

(6) **Operation security.** Operation security includes deception, physical security, signal security (SIGSEC), and information security. It is taught in all training and at all levels. Although all areas of OPSEC must be measured in training, techniques of camouflage and SIGSEC need special emphasis. (See AR 530-1 and TC 100-1 for more information.)

G-10. Professional development.

Ranger units at battalion and regimental levels conduct many officer and NCO professional development programs. This ensures that all officers and NCOs are skilled in their field, confident, and wellrounded. Professional readings and book reports, tactics seminars, and tactical exercises without troops are good development methods.

G-11. Physical training.

a. This program ensures that each ranger maintains the physical ability and stamina needed to do his mission. Other objectives are to promote unit cohesiveness and discipline. These are met by combining individual testing, formation, running, athletic competition, weight control, proper diet, teamwork drills, and freedom from drug and alcohol abuse. Mental fitness and toughness come from experience and success. Good physical conditioning increases a unit's chances of success.

b. Physical training (PT) programs are designed to include the elements of overload, progression, balance, variety, and regularity.

c. Ranger unit physical fitness standards are:

(1) Score 80 points for each Army physical readiness test (APRT) event and do six chinups.

(2) Pass the ranger swimming test.

(3) Complete an 8-kilometer run in 40 minutes.

(4) Complete a 12-kilometer road march in three hours (with rucksack, helmet, and weapon).

(5) Meet the Army height and weight standards.

G-12. Environmental and specialized training.

a. Rangers conduct training in environments as close to real combat missions as possible. This conditions rangers to survive and fight in a specific environment. Such training includes combat-critical tasks and technique training for that environment.

b. Each ranger battalion's environmental and specialized training includes at least the following:

(1) Jungle (once every 12 months).

(2) Desert (once every 12 months).

(3) Urban (once every 6 months).

(4) Extreme cold (once every 18 months).

(5) Mountain (once every 12 months).

(6) Amphibious (once every 18 months).

G-13. Ranger indoctrination program.

a. The RIP is a training and selection program. It is run by specially selected NCOs from the regimental headquarters. It has as its mission the training and indoctrination of newly assigned soldiers. It physically toughens and mentally sharpens the newly assigned soldier. It teaches basic skills and techniques. The RIP identifies and eliminates in three weeks any applicant who does not show dedication, motivation, physical fitness, and emotional stability.

b. An indoctrination program for newly assigned ranger-qualified officers and NCOs prepares them to assume duty as leaders within the ranger regiment. These personnel monitor and participate in RIP training, as well as undergoing individual ranger leadership training. The RIP provides them an overview of the regiment's history, its mission, and its goals. (See Appendixes C,

D, and E for further information about the history and goals of the ranger regiment.) The RIP also provides a review of the skills required in a ranger unit. Refresher training in airborne and pathfinder operations, use of indirect fires, and light infantry doctrine is conducted. During RIP all newly assigned officers and NCOs must meet the ranger physical fitness standard.

c. The RIP cadre also conducts a preranger program to prepare soldiers to attend the US Army Ranger School. It is a physically demanding course that teaches troop-leading procedures, operation orders, and small unit patrol actions.

Glossary

The following acronyms, abbreviations, and definitions are provided to aid the reader in understanding this publication.

Section I

ACRONYMS AND ABBREVIATIONS

AAT Army assault team

ABCCC airborne battlefield command and control center

ACL allowable cargo load

ADA air defense artillery

AFFOR the USAF element of a joint task force

AGL above ground level

AKZ armor kill zone

ALCC airlift control center

ALCE airlift control element

AM amplitude modulation

AMC air mission commander

APERS antipersonnel

APRT Army physical readiness test

ARFOR the Army element of a joint task force

ARSOC Army special operations command

ARSOF Army Special Operations Forces

ASG area support group

ASPC all-source production center

ATGM	antitank guided missile
AWADS	adverse weather aerial delivery system
BCE	battle coordination element
BTMS	battalion training management system
CA	civil affairs
CAS	close air support
CCT	combat control team
CDS	container delivery system
CE	communications-electronics
CI	counterintelligence
CINC	commander in chief
COMINT	communications intelligence
COMSEC	communications security
CONUS	continental United States
COSCOM	corps support command
CP	command post
CS	combat support; a chemical agent
CSS	combat service support
DACG	departure airfield control group
DACO	departure airfield control officer
DASC	direct air support center
DMDG	digital message device group
DOD	Department of Defense
DS	direct support
DSU	direct support unit
DZ	drop zone
EAC	echelons above corps

EACIC	Echelons Above Corps Intelligence Center
ECCM	electronic counter-countermeasure
ECM	electronic countermeasures
E&E	escape and evasion
EEFI	essential elements of friendly information
EOC	emergency operations center
EW	electronic warfare
FARP	forward area rearm and refuel point
FASCAM	family of scatterable mines
FAX	facsimile transmission
FDC	fire direction center
FEBA	forward edge of the battle area
FFL	friendly front lines
FIST	fire support team
FLOT	forward line of own troops
FM	field manual; frequency modulation
FO	forward observer
FORSCOM	US Army Forces Command
FPF	final protective fire
FPL	final protective line
FRAGO	fragmentary order
FSA	fire support area
FSCOORD	fire support coordination or coordinator
FSE	fire support element
FSO	fire support officer
FSS	fire support station
FTX	field training exercise

G3	Assistant Chief of Staff, Operations
GS	general support
GSR	ground surveillance radar
HAARS	high-altitude, airdrop resupply system
HAHO	high-altitude, high-opening
HALO	high-altitude, low-opening
HE	high explosive
HF	high frequency
HHC	headquarters and headquarters company
HSLLADS	high-speed, low-level airdrop system
HUMINT	human intelligence
IAW	in accordance with
IFF	identification-friend-or-foe
IP	initial point
IRP	initial rally point
ISB	intermediate staging base
JACC/CP	joint airborne communications center/command post
JCS	Joint Chiefs of Staff
JCSE	joint communications support element
JFC	joint force commander
JSOC	joint special operations command
JTB	joint transportation board
JTF	joint task force
kw	kilowatt
LAPES	low altitude parachute extraction system
LAW	light antitank weapon
LNO	liaison officer

LOC	logistical operations center
LRRP	long-range reconnaissance patrol
LRSU	long-range surveillance unit
LTD	laser target designator
LZ	landing zone
MAC	Military Airlift Command
MANPADS	man-portable air defense system
MCC	movement control center
METT-T	mission, enemy, terrain, troops and time available
MI	military intelligence
MILES	multiple integrated laser engagement system
mm	millimeter
MMC	Materiel Management Center
MMEE	minimum mission-essential equipment
MOS	military occupational specialty
MOUT	military operations on urbanized terrain
MRLS	multiple rocket launch system
MTP	mission training plan
NBC	nuclear, biological, chemical
NCA	National Command Authority
NCO	noncommissioned officer
NOD	night observation device
OCOKA	observation, concealment, obstacles, key terrain, and avenues of approach
OCONUS	outside continental United States
OPCOM	operational command
OPCON	operational control
OPFOR	opposing force

OPORD	operation order
OPSEC	operations security
ORP	objective rally point
PAC	personnel administration center
PDF	primary direction of fire
PIR	priority intelligence requirements
PLL	prescribed load list
POM	preparation for oversea movement
PSYOP	psychological operations
PT	physical training
PZ	pickup zone
QRE	quick-reaction element
RATELO	radiotelephone operator
REMAB	remote marshalling base
RFL	restrictive fire line
RIP	ranger indoctrination program
RP	rally point; release point
RPG	rocket-propelled grenade
RRF	ranger ready force
RRP	reentry rally point
RSE	ranger support element
RSTA	reconnaissance, surveillance, and target acquisition
S1	adjutant
S2	intelligence officer
S3	operations and training officer
S4	supply officer
S5	civil affairs officer

SAR search and rescue

SATCOM satellite communications)

SCIF sensitive compartmented information facility

SCOPES squad combat operations' exercises (simulation)

SCUBA self-contained underwater breathing apparatus

SEAD suppression of enemy air defense

SEMA special electronic mission aircraft

SERE survival, evasion, resistance, and escape

SF special forces

SHORAD short-range air defense

SIGSEC signal security

SKE station-keeping equipment

SLAR side-looking airborne radar

SOA special operations aviation

SOCB special operations communications battalion

SOCOM special operations command (usually, 1st SOCOM)

SOP standing operating procedures

SOSB special operations support battalion

SOTF special operations task force

SPL support platoon leader

STX situational training exercise

TA theater Army

TAACOM Theater Army Area Command

TAC Tactical Air Command

TACP tactical air control party

TACSAT tactical satellite terminal

TAMMC Theater Army Materiel Management Center

TASG	theater Army support group
TDA	table of distribution and allowances
TOA	transfer of authority
TOC	tactical operations center
TOE	table(s) of organization and equipment
TRP	target reference point
TTY	teletype
UHF	ultrahigh frequency
US	United States
USAF	United States Air Force
USMC	United States Marine Corps
USN	United States Navy
VHF	very high frequency
WP	white phosphorus
XO	executive officer

Section II

DEFINITIONS

ADVANCE FORCE. An operation requiring the immediate deployment of US forces with the intent to negate a threat before the deployment of follow-on forces is required.

AIRLAND BATTLE DOCTRINE. An approach to military operations that realizes the full potential of US forces. Two notions (extending the battlefield and the ability to integrate conventional, nuclear, chemical, and electronics means) are combined to describe a battlefield where the enemy is attacked to the full depth of his formation.

AREA OF INFLUENCE. A geographical area wherein a commander is directly capable of influencing operations by maneuver or fire support systems normally under his command or control.

AREA OF INTEREST. That area of concern to the commander, including the area of influence, areas adjacent thereto, and extending into enemy territory to the objectives of current or planned operations. This area also includes areas occupied by enemy forces who would jeopardize the accomplishment of the mission.

AREA OF OPERATIONS. That portion of an area of conflict necessary for military operations. Areas of operations are geographical areas assigned to commanders for which they have responsibility and in which they have authority to conduct military operations.

CHALK. One specific aircraft load.

CLANDESTINE OPERATIONS. Activities to accomplish intelligence, counterintelligence, and other similar activities sponsored or conducted by governmental departments or agencies, in such a way as to ensure secrecy or concealment. (They differ from covert operations in that emphasis is placed on concealment of the operation rather than on concealment of the identity of the sponsor.)

COMMUNICATIONS INTELLIGENCE (COMINT). The interception and processing of foreign communications passed by radio, wire, or other electromagnetic means, and by the processing of foreign encrypted communications, however transmitted. Interception comprises search, intercept, operator identification, signal analysis, traffic analysis, crypto-analysis, study of plain text, the fusion of these processes, and reporting the results.

COMMUNICATIONS SECURITY. The protection resulting from all measures designed to deny to unauthorized persons information of value that might be derived from the possession and study of telecommunications, or to mislead unauthorized persons in their interpretations of the results of such study. COMSEC includes crypto-security, physical security, and transmission security.

COUNTERINTELLIGENCE. That aspect of intelligence activities, both offensive and defensive, designed to detect and neutralize or destroy the effectiveness of hostile foreign intelligence activities and to protect information against espionage, personnel against subversion, and installations and materiel against sabotage.

COVERT OPERATIONS. Operations that are so planned and executed as to conceal the identity of or permit plausible denial by the sponsor. They differ from clandestine operations in that emphasis is placed on concealment of identity of the sponsor rather than on concealment of the operation.

DECEPTION. Those measures designed to mislead the enemy by manipulation, distortion, or falsification of evidence to induce him to react in a manner prejudicial to his interests.

DEMONSTRATION. A show of force in an area where a decision is not sought. It is similar to a feint but does not make contact with the enemy.

DRILL. A drill focuses on a "chunk" or "slice" of battle and is targeted at small units. Drills are critical collective tasks that require a high degree of proficiency from a small unit. Generally, small units "drill" collective tasks that require rapid responses by unit members in the absence of detailed orders from unit leaders.

ELECTRONIC COUNTERMEASURES. That division of electronic warfare involving actions taken to prevent or reduce an enemy's effective use of the electromagnetic spectrum.

ELECTRONIC INTELLIGENCE (ELINT). The technical and intelligence information derived from foreign noncommunications electromagnetic radiations emanating from other than atomic detonations or radioactive sources.

ELECTRONIC WARFARE. Military action involving the use of electromagnetic energy to determine, exploit, reduce, or prevent hostile use of the electromagnetic spectrum and action that retains friendly use of the electromagnetic spectrum.

ELECTRONIC WARFARE SUPPORT MEASURES (ESM). That division of EW involving actions taken to search for, intercept, identify, and locate sources of radiated electromagnetic energy for the purpose of immediate threat recognition. ESM provides a source of information needed for immediate action involving electronic countermeasures, electronic counter-countermeasures, avoidance, targeting, and other tactical employment of forces.

ESPIONAGE. Actions directed toward the acquisition of information through clandestine operations.

EVACUATION/SAFEGUARD. The use of US forces to evacuate US citizens or allies from a hostile or potentially hostile area, or to safeguard US personnel or property.

EXFILTRATION. The removal of personnel or units from areas under enemy control by stealth, deception, surprise, or clandestine means. Exfiltration is a method of extracting ranger units.

EXTRACTION. The removal of personnel or units from areas under enemy control by any means, to include exfiltration, airland, airmobile, or amphibious.

FEINT. A diversionary supporting attack conducted to draw the enemy's attention from the main effort. It is normally executed by brigades and small units. Feints are usually shallow, limited objective attacks that go in before or during the main attack.

GUERRILLA WARFARE. Military and paramilitary operations conducted in enemy-held or hostile territory by irregular and predominantly indigenous forces.

HIGH-INTENSITY CONFLICT. A war between the US and any other nation or nations in which both belligerents use the most modern technology and all resources in intelligence, mobility, firepower (including NBC), command, communications and control, and service support.

HUMAN INTELLIGENCE. Intelligence derived from information provided by human resources.

INFILTRATION. The movement through or into an area of territory occupied by either friendly or enemy troops or organizations. The movement is made either by small groups or by individuals, at extended or irregular intervals. When used in conjunction with the enemy, it implies that contact is avoided. Infiltration is a subset of insertion.

INSERTION. Placement of troops and equipment into an operational area by any means.

INTERDICT. To isolate or seal off an area by military means; to prevent, hinder, or delay the use of an area or route by enemy forces.

INTERDICT LINE OF COMMUNICATIONS. An attack to seal off an area or to deny use of a route or approach.

JOINT TASK FORCE. A JTF may be constituted and designated by the Secretary of Defense or by the commander of a unified command, specified command, or an existing JTF. Normally, it performs missions having specific, limited objectives or missions of short duration. It dissolves when it has achieved its purpose. The joint task force commander is responsible to the commander, known as the establishing authority, who created the JTF. The JTF is composed of elements of two or more services operating under a single JTF commander. The JTF commander has operational control over the entire force and may have direct command of his own staff with representatives from the other services. He exercises logistics coordination or control only as necessary to meet his subordinate commanders' logistics needs.

LIAISON. That personal contact or communication maintained between elements of military forces to ensure mutual understanding and unity of purpose and effort.

LINE OF COMMUNICATIONS. All routes (land, water, and air) that connect an operating military force with one or more bases of operations and along which supplies and military forces move.

LOW-INTENSITY CONFLICT. Limited politico-military operations conducted to achieve political, social, economic, or psychological objectives. It is generally confined to a specific geographic area and is often characterized by operational constraints imposed on ranger units by political and economic considerations. It may involve intermittent combat against regular threat military forces, but it more often involves conflict with irregular, security, or paramilitary forces of a country or group involved in an operation that is counter to the best interests and national security of the United States. It is characterized by relatively low levels of combat and constraints placed on the weaponry, tactics, and movements of the ranger force. It may involve counterterrorist or antiterrorist operations and generally occurs before a pro forma declaration of war.

MID-INTENSITY CONFLICT. War, declared or undeclared, between the US and the regular forces of one or more nations and their respective allies, if any, in which the belligerents use the most modern technology and all resources in intelligence, mobility, firepower (excluding nuclear, chemical, and biological weapons), command, communications and control, and support. They are used for limited objectives under definitive policy limitations as to the extent of destructive power that can be used or the extent of geographical area that might be involved.

MILITARY STRATEGY. Art and science of using the armed forces of a nation to secure the objectives of national policy by application of force or the threat of force. Military strategy sets the fundamental conditions for operations.

NATIONAL COMMAND AUTHORITY. This includes the President and Secretary of Defense or the duly deputized alternates or successors. Commonly referred to as NCA.

OPERATIONAL COMMAND/CONTROL. Operational command (OPCOM) and operational control in joint force terminology both refer to the authority exercised by joint commanders over subordinate service components. Those terms are not interchangeable in joint operations. OPCOM applies to the authority exercised by commanders of unified commands. OPCON is the authority that subordinate joint task force commanders exercise in the conduct of specific operations. The authority that military departments exercise over their respective components is commonly referred to as command less operational command.

OPERATIONAL LEVEL OF WAR. The operational level of war uses available military resources to attain strategic goals within a theater of war. It is the theory of larger unit operations. It also involves planning and conducting campaigns. Campaigns are sustained operations designed to defeat an enemy force in a specified space and time with simultaneous and sequential battles. The disposition of forces, selection of objectives, and actions taken to weaken or to outmaneuver the enemy all set the terms of the next battle and exploit tactical gains. They are all part of the operational level of war. In AirLand Battle doctrine, this level includes the marshalling of forces and logistical support, providing direction to ground and air maneuver, applying conventional and nuclear fires in depth, and employing unconventional and psychological warfare.

OPERATIONS SECURITY. Those measures designed to protect information concerning planned, ongoing, and completed operations from unauthorized disclosure. It includes all actions a command takes to deny the enemy information about friendly units and their operations.

ORDER OF BATTLE (OB). The identification, strength, command structure, and disposition of the personnel, units, and equipment of any military force.

RAID. Usually a small-scale operation, involving a swift penetration of hostile territory to secure information, confuse the enemy, or destroy his installations. It ends with a planned withdrawal upon completion of the assigned mission.

RANGER MISSION. To conduct strike, raid, or special light infantry operations. Execution of ranger missions is normally directed by the National Command Authority. Rangers also support operational-level commanders during operations of limited duration when allocated or assigned by the theater commander.

RECONNAISSANCE. A mission undertaken to obtain, by visual observation or other detection methods, information about the activities and resources of an enemy or potential enemy; or to secure data concerning the meteorological, hydrographical, or geographical characteristics of a particular area.

SEARCH AND RESCUE. The use of aircraft, surface craft, submarines, specialized rescue teams, and equipment to search for and rescue personnel in distress on land or at sea.

SEIZURE. As an operational purpose, the capture of a voluntarily restricted portion of an enemy-controlled territory. Capture of an isolated land mass such as an island is usually categorized as a seizure.

SEIZE KEY FACILITY. An attack on an installation with the intent of holding it for a short period. Normally, rangers would make the assault to seize and secure an installation for follow-on forces.

SERVICE COMPONENT. Each service component commander is responsible for recommending the proper use of his forces and for accomplishing operational tasks assigned by the joint commander. He is also responsible for his service in matters of internal administration and discipline; training in own service doctrine, techniques, and tactics; designation of specific units to meet joint requirements; logistics functions normal to the component; tactical employment of service component forces; and service intelligence matters.

SHOW OF FORCE. A mission carried out to demonstrate US resolve, whereby US forces are deployed to defuse a situation that may be harmful to US interests or national objectives.

SIGNAL INTELLIGENCE (SIGINT). A generic term that includes COMINT and ELINT. Activities concerned with the collection and processing of foreign electromagnetic emissions for the purpose of producing intelligence information.

SITUATIONAL TRAINING EXERCISE. An STX is a short, mission-oriented exercise in which a group of closely-related collective tasks are trained, using a doctrinally preferred method of executing these tasks to established standards. An STX is drill-like in nature, but it is more complex and flexible than a drill. It usually involves a larger unit than a drill.

SPECIAL OPERATIONS. Military operations conducted by specially trained, equipped, and organized DOD forces against strategic or tactical targets in pursuit of national, military, political, economic, or psychological objectives. Special operations may support conventional military operations, or they may occur independently when the use of conventional forces is either inappropriate or infeasible. Sensitive peacetime operations, except for training, are normally authorized by the NCA and conducted under the direction of the NCA or designated commander. Special operations may include unconventional warfare, counter-terrorist operations, collective security, psychological operations, and civil affairs measures.

SPECIAL OPERATIONS FORCES. All civil affairs, psychological operations, ranger (active component), and special forces units within the US Army, US Army Reserve, and National Guard.

SPECIAL OPERATIONS COMMAND. The major subordinate command to FORSCOM exercising command and control of all active component Army special operations forces not placed under the operational command of unified or specified commands.

STRATEGIC LEVEL OF WAR. Military strategy uses the armed forces of a nation to secure the objectives of national policy by applying force or the threat of force. Military strategy sets the fundamental conditions for operations.

STRIKE. An attack that is intended to inflict damage on, seize, or destroy an objective. Sometimes referred to as direct action missions.

SURVEILLANCE. The systematic observation of aerospace, surface or subsurface areas, places, persons, or things by visual, aural, electronic, photographic, or other means.

TACTICAL. Pertaining to the employment of units in combat.

TACTICAL LEVEL OF WAR. Tactics are the specific techniques smaller units use to win battles and engagements that support operational objectives. Tactics employ all available combat, combat support, and combat service support. Tactics involve the movement and positioning of forces on the battlefield in relation to the enemy, the provisioning of fire support, and the logistical support of force before, during, and following engagements with the enemy. At corps and division, operational and tactical levels are not clearly separable, but they are guided by the same principles. An operation designed to defeat any enemy force in an extended area does so through operational maneuver and a series of tactical actions.

TACTICAL COMBAT INTELLIGENCE. That knowledge of the enemy, weather, and geographical features needed by a commander in the planning and conduct of combat operations. It is derived from the interpretation of information on the enemy and the environment.

TARGET ANALYSIS. A detailed and systematic examination of processed intelligence to identify and locate targets.

UNIFIED COMMAND. A unified command is established and designated by direction of the President. The command is organized under a single unified CINC and has a broad, continuing mission. The CINC is responsible to the Secretary of Defense for accomplishing his broad and continuing mission and operates under the strategic and operational direction of the JCS. A unified command is composed of assigned forces of two or more services. These forces, as service components, are under the operational command of the unified commander. The CINC exercises operational command authority through service component commanders. When tasked by higher authority, he may establish a subordinate JTF to conduct specific missions. Under emergency conditions, the CINC has the authority to use all facilities and supplies of assigned forces to accomplish his mission.

References

REQUIRED PUBLICATIONS

Required publications are sources that users must read in order to understand or to comply with this publication.

Field Manuals (FM)

FM 7-8	The Infantry Platoon and Squad (Infantry, Airborne, Air Assault, Ranger)
FM 7-10	The Infantry Rifle Company (Infantry, Airborne, Air Assault, Ranger)
FM 7-20	The Infantry Battalion (Infantry, Airborne, and Air Assault)

RELATED PUBLICATIONS

Related publications are sources of additional information. They are not required in order to understand this publication.

Army Regulations (AR)

AR 350-1	Army Training

Field Manuals (FM)

FM 10-52-1	Commander's Handbook for Water Usage in Desert Operations
FM 10-501	Airdrop of Supplies and Equipment: Rigging Containers
FM 10-512	Airdrop of Supplies and Equipment: Rigging Typical Supply Loads
FM 10-542	Airdrop of Supplies and Equipment: Rigging Loads for Special Operations
FM 10-547	Airdrop of Supplies and Equipment: Rigging the High Speed Aerial Delivery Container CTU-2/A
FM 10-553	Airdrop of Supplies and Equipment: Rigging Ammunition
FM 12-3-4	Echelons Above Corps (EAC) Personnel and Administrative Doctrine

FM 14-6	Comptroller/Finance Services in Theaters of Operations
FM 19-15	Civil Disturbances
FM 20-11-1	Military Diving (Volume 1)
FM 20-11-2	Military Diving (Volume 2)
FM 21-20	Physical Training Program
FM 21-26	Map Reading
FM 21-75	Combat Skills of the Soldier
FM 21-76	Survival, Evasion, and Escape
FM 21-77A(S)	Joint Worldwide Evasion and Escape Manual
FM 22-9	Soldier Performance in Continuous Operations
FM 23-67	Machinegun, 7.62-mm, M60
FM 25-1	Training
FM 25-2	Unit Training Management
FM 25-3	Training in Units
FM 25-4	How to Conduct Training Exercises
FM 25-5	Training for Mobilization and War
FM 25-7	Training Ranges
FM 26-2	Management of Stress in Army Operations
FM 29-47	Main Supply and Service Company, Forward Supply and Service Detachments, Supply and Service Battalion, Airborne Division
FM 29-51	Division Supply and Field Service Operations
FM 30-5	Combat Intelligence
FM 30-17	Counterintelligence Operations
FM 31-18	Long-Range Reconnaissance Ranger Company
FM 31-19	Special Forces Military Free-Fall Parachuting
FM 1-24	Special Forces Air Operations

FM 31-25	Special Forces Waterborne Operations
FM 31-70	Basic Cold Weather Manual
FM 34-1	Intelligence and Electronic Warfare Operations
FM 34-60A	Counterintelligence Operations (U)
FM 55-9	Unit Air Movement Planning
FM 57-38	Pathfinder Operations
FM 57-220	Basic Parachuting Techniques and Training
FM 63-3J	Combat Service Support Operations--Corps
FM 63-4	Combat Service Support Operations--Theater Army Area Command
FM 63-5	Combat Service Support Operations--Theater Army
FM 90-3	Desert Operations
FM 90-4	Airmobile Operations
FM 90-5	Jungle Operations
FM 90-6	Mountain Operations
FM 90-8	Counterguerilla Operations
FM 90-10	Military Operations on Urbanized Terrain (MOUT)
FM 90-10-1	An Infantryman's Guide to Urban Combat
FM 100-5	Operations
FM 100-20	Low Intensity Conflict
FM 100-26	The Air-Ground Operations System
FM 101-5	Staff Organization and Operations
FM 101-10-1	Staff Officers' Field Manual: Organization, Technical and Logistical Data (U)

Firing Table (FT)

FT 60-P-1	Firing Tables for Mortar, 60-mm: M224, Firing Cartridge, HE, M720, Cartridge, HE, M49A4, Cartridge, TP, M5OA3, Cartridge, WP, M302AI, Cartridge, Illuminating, M83A3

Soldier's Training Publication (STP)

STP 7-11B24-SM	Soldier's Manual MOS 11B, Infantryman, Skill Level 2/3/4

Technical Manuals (TM)

TM 10-1670-267-12&P	Operator's and Organizational Maintenance Manual Including Repair Parts and Special Tools List for High Altitude Airdrop Resupply System: 500 Pound Capacity

Training Circulars (TC)

TC 23-14	Sniper Training and Employment
TC 23-90	Mortar Training
TC 24-1	Communications-Electronics Operation Instructions, the CEOI
TC 31-20-5(C)	Special Forces Communications (U)
TC 90-6-1	Military Mountaineering

Other Related Publications

King, Michael J. Dr., *Leavenworth Papers, No. 11*, "Rangers Selected Combat Operations in World War II," Combat Studies Institute, USACGSC, Fort Leavenworth, KS

PROJECTED PUBLICATIONS

Projected publications are sources of additional information that are scheduled for printing but are not yet available. Upon print, they will be distributed automatically via pinpoint distribution. They may not be obtained from the USA AG Publications Center until indexed in DA Pamphlet 310-1.

Field Manuals (FM)

FM 7-70	Light Infantry Squad/Platoon
FM 7-71	Light Infantry Company
FM 7-72	Light Infantry Battalion
FM 7-73	Light Infantry Brigade

FM 23-36 60-mm Mortar, M224

Made in United States
North Haven, CT
04 November 2026

21773562R00178